Farmer Participation
and Irrigation Organization

Studies in Water Policy and Management
Charles W. Howe, General Editor

Studies in Water Policy and Management

Charles W. Howe, General Editor

Farmer Participation and Irrigation Organization, edited by Bradley W. Parlin and Mark W. Lusk

Water Resources: Distribution, Use, and Management, John R. Vincent and R. Liu

Social, Economic, and Institutional Issues in Third World Irrigation Management, edited by R. K. Sampath and Robert A. Young

Irrigation Management in Developing Countries: Current Issues and Approaches, edited by Jon P. Brooks and Derrick J. Thom

Congress in Its Wisdom: The Bureau of Reclamation and the Public Interest, Doris Ostrander Dawdy

Water Marketing and the Law: Above-Ground Water Markets and Public Policy Issues, edited by David B. Bush

Instream Flow Protection: Seeking a Balance with Current Appropriation Doctrine, the Prior-Right Defense, Marcia J. Upsall

Irrigation Investment, Technology, and Management Strategies for Development, edited by K. William Easter

Irrigation Management in Developing Countries: Current Issues and Approaches, edited by K. C. Nobe and R. K. Sampath

Farmer Participation
and
Irrigation Organization

edited by
Bradley W. Parlin
and
Mark W. Lusk

Routledge
Taylor & Francis Group
LONDON AND NEW YORK

To Dr. Howard D. Lusk

First published 1991 by Westview Press

Published 2018 by Routledge
52 Vanderbilt Avenue, New York, NY 10017
2 Park Square, Milton Park, Abingdon, Oxon OX14 4RN

Routledge is an imprint of the Taylor & Francis Group, an informa business

Library of Congress Cataloging-in-Publication Data
Farmer participation and irrigation organization / edited by Bradley
    W. Parlin and Mark W. Lusk.
        p. cm.—(Westview studies in water policy and
management)
    Includes bibliographical references.
    ISBN 0-8133-8324-2
    1. Irrigation—Management—Citizen participation.   2. Farmers.
3. Irrigation—Developing countries—Management—Citizen
participation—Case studies.   I. Parlin, Bradley W.   II. Lusk, Mark
W., 1943– .   III. Series.
TC812.F37   1991
33.91'315—dc20   91-3709
                                                                                            CIP

ISBN 13: 978-0-367-01961-7 (hbk)
ISBN 13: 978-0-367-16948-0 (pbk)

Contents

Tables and Figures

Figures

Tables

Foreword

It has been a joy for me to review this volume covering the social aspects of improving the performance of publicly developed and managed irrigation schemes. The text is not only directed at the importance of farmer participation, but also at the equal importance of bureaucratic design and management necessary for scheme success.

The editors start with a convincing argument for greater application of social science research--research that makes better use of the implications of theoretically driven, bureaucratic design and management strategies. They point out reasons why the application of sociology to irrigation development and management produces its best yields when grounded by relevant theoretical perspectives.

They review the advantages and limitations of applying organizational, human ecology, conflict, and public (or rational) choice sociological theories to irrigation scheme management. The use of public choice theory is singled out as being particularly relevant, although not heretofore utilized, for irrigation management studies. This is because the theory is a welding of economics, political science, and sociology as they apply to decision-making.

The text continues by expanding on the various aspects of the social science framework's relevancy and application to irrigation development and management. This is followed by a series of case studies (mostly focusing on farmer participation) from throughout the major irrigated areas of the world.

The book does an excellent job of stressing the fact that the function of irrigation organization is to design and manage institutions and physical structures that economically deliver water in a timely and reliable manner with the highest practical degree of control at the farm level. Thus, development schemes or on-farm water management projects are never solely a matter of changing individual farmer behavior, but are also based on building the connections between farmers and between farmers and organizations. By using the rational choice model to load the development equation on behalf of success, the editors transcend the usual problem solving or *clinical* framework. The clinical approach seeks to resolve project difficulties on an atomistic, post-hoc basis using specialists to *diagnose* project ills.

Their system approach gives equal or greater emphasis to project performance as opposed to project problems. It incorporates development inputs, incentives, counterdevelopment forces, environmental factors, and feedback to provide a holistic cortex for irrigation project management.

xi

The editors of this volume have captured an important dimension of the social science implications for the successful organization and management of large scale public irrigation projects. Their new insights regarding the organization of water management should bear fruitful results.

As an irrigation engineer, I worked closely with the editors in an interdisciplinary effort to articulate and improve our ability to transfer our collective irrigation management expertise. This was done as the Irrigation Experience Transfer sub-activity of the United States Agency for International Development Water Management Synthesis II Projects.

Jack Keller
Professor Emeritus
Department of Agricultural
and Irrigation Engineering
Utah State University

Preface

The challenge of irrigation development is partly rooted in the complex character of the systems which are being targeted for social and technical change. Irrigation systems incorporate at least three major domains--the watershed, conveyance, and agricultural. It is difficult enough to design irrigation projects that effectively integrate these domains, but irrigation development is made enormously more complicated by virtue of the fact that it involves changes in a fourth socio-cultural domain and therefore requires the mobilization and coordination of numbers of people for it to succeed. The socio-cultural domain in irrigation development is the topic of this book. This development realm consists of the organizational and institutional elements of an irrigation system which are affected by human values, laws, policies, skills, and technologies.

The purpose of this book is to introduce and illustrate fundamental knowledge in the social sciences relevant to irrigation organization and farmer participation in order to facilitate a more comprehensive understanding of irrigation development and water management.

Part One introduces the elements of a social scientific understanding of irrigation development. Chapter One reviews theoretical orientations that have been used in understanding water management with emphasis on the importance of economic factors and rational choice in current theory. Chapter Two discusses the design of middle level organizations for managing the interface between farmers and irrigation bureaucracies. Chapter Three summarizes key lessons learned by social scientists in irrigation project experience.

In Part Two, authors from around the world review country and regional case examples of attempts to develop effective irrigation organizations and induce farmer participation. These cases reflect a diverse spectrum of cultures and national contexts in order to demonstrate and apply the basic concepts of irrigation organization and farmer participation in varied settings. Examples from Asia, North America, South America and Africa are included. Country and region cases were selected to demonstrate successes associated with democratization, privatization and decentralization as well as to show how systems that do not incorporate these elements are more subject to disorganization and breakdown.

Bradley W. Parlin and Mark W. Lusk

PART ONE

Farmer Participation and Irrigation Organization: An Overview

1

Bureaucratic and Farmer Participation in Irrigation Development

Mark W. Lusk and Bradley W. Parlin

The development of intervention strategies designed to effectively involve farmers in the management of international irrigation projects has become the focus of considerable research and experimentation by social scientists (Parlin & Lusk, 1988; Tirmizi & Parlin, 1990). Indeed, the literature on various aspects of farmer participation in irrigation development is extensive. Studies of water user associations (WUAs) worldwide have contributed to the understanding of the costs, benefits and situational determinants which impede or facilitate farmer involvement in irrigation organization (FAO, 1985; Uphoff, Meinzen-Dick, & St. Julien, 1985). Despite the substantial knowledge base concerning farmer participation in irrigation projects, intervention efforts have met with mixed reviews. In some cases organization strategies have met with modest success and others with dismal failure. In Sri Lanka, for example, the lessons learned in the National Irrigation Association experiments in the Philippines were successfully transferred to the rehabilitation of the left bank of the Gal Oya system (Wijayaratne, 1984; Uphoff et al., 1985). However, efforts to organize farmers in the Mahaweli Development scheme have been fraught with difficulties such as sabotage, deviant irrigation practices, settler apathy, and disengagement (Scudder, 1985; Nott, 1985; Parlin, Lusk, & Al-Rashid, 1986). A central question this book addresses is why attempts to organize and involve farmers in irrigation management have led to such variable outcomes, especially given the well-documented results of scholarly and applied efforts in this regard.

The premise of this book is that social science research on irrigation development has given insufficient attention to the implications of theoretically-driven, bureaucratic design and management strategies for project success. While social science research has provided substantial insight into the motivational basis of farmer participation, there has been little corresponding research on bureaucratic behavior and its implications for farmers in development settings. Given that farmers and bureaucrats comprise the two principal actors in the irrigation drama, it would appear that social science research has focused upon changing farmer behavior while ignoring bureaucratic behavior and its impact.

The social science literature on irrigation development has generally been characterized by site specific descriptive, or analytical case level studies (Illo & Chiong-Javier, 1983; Ganewatte, 1985). Importantly, most studies of farmer participation based on field research experiences have not been based upon theory. A theoretical framework to facilitate the integration of the diverse wealth of case level findings would provide a useful tool to systematically guide informed inquiry into the sociology of irrigation development. A major purpose of this chapter is to add to our understanding of the irrigation development process by proposing a model of irrigation development behavior utilizing rational choice theory.

Bureaucracy and Irrigation Management

Despite the rich tradition of research in the sociology of complex organizations (associations, bureaucracies), few writers have attempted to apply the theoretical, conceptual and substantive knowledge of the field to irrigation management and development. This important gap has been noted by Borlaug (1987:387), for example, who has emphasized the limits on efficiency of third world irrigation schemes posed by *a bloated bureaucracy.* Bromley (1987:173) stresses the need for research on irrigation organization in order to overcome the *institutional vacuum* which characterizes the management of irrigation programs in the developing world. Wade (1987:198) also sees organizational reform as a central issue in irrigation development. He argues persuasively that the failure of many systems is sometimes a result of physical design but that, "part of the answer is to be found in the design of the irrigation management organization" (1987:178). Another advocate of organizational reform in irrigation management is Chambers (1987). He calls for a *professional revolution* in which irrigation bureaucrats, following the lead of successful U.S. private sector industrial organizations, become responsive to the needs of their clients (farmers), and thereby contravene what Hart (1978) describes as a *syndrome of anarchy* in irrigation management.

The importance of irrigation to rural development and the potential for social conflict, corruption and disorganization make irrigated agriculture an ideal context for assessing applications of the sociology of organizations. A goal of this text is to develop a general strategy that stresses farmer - constituency incentives in the building of organizations as an alternative to focusing on the traditionally identified obstacles to irrigation development. Such an approach would vary from past practice by emphasizing positive opportunities in water management rather than problems and diagnoses (Keller, 1988).

Irrigation Development: Theoretical Perspectives

The relative absence of attention to the organizational and management side of irrigation development emerges, in part, from the lack of theoretical underpinnings in the case level studies which have provided most of the knowledge base in irrigation development.

Theoretical sociological thinking with regard to irrigation is nascent. Few theoreticians have written on the subject of irrigated societies or communities. As Bromley (1982:3) notes, "while there is extensive theoretical literature on landlord-tenant relations, we do not have anything comparable in irrigated agriculture." This trend is unfortunate in light of the fundamental importance of irrigated agriculture to many societies in the contemporary world and throughout history. In addition, much of the current writing on irrigation development and management by social scientists is at the descriptive level. While the use of social science in irrigation development and management is necessarily an applied endeavor, a key purpose of social theory has been to guide such social practice.

We will review a handful of theoretical approaches that have some relevance for understanding human behavior in irrigation communities. Our choice of theories to review is not exhaustive, but rather illustrates the utility of sociological theory for clarifying the process of irrigation development.

Development Theory

Sociology and anthropology have addressed the challenge of development and planned social/technological change using sociological and political theory. A rich literature has dealt with the implications of modernization (Rostow, 1971), neo-colonialism and dependency (Frank, 1969), economic growth (Hagen, 1968), and technological innovation (Spicer, 1967) to the development process. Although important to a general understanding of societal evolution, these classical perspectives have had limited application to the specific problems of local projects.

Growing out of these traditions is the theme that development is the process of overcoming obstacles to planned social change. Depending upon the theoretical perspective, the obstacles identified differ significantly. Modernization theorists like Gunnar Mydral (1963) would emphasize internal obstacles such as the lack of *entrepreneurship*. Others within the school might focus on the critical importance of culture (Banfield, 1958). An extreme form of the modernization approach stresses the psychological attributes of peoples and characterizes underdevelopment as a *state of mind* (Harrison, 1985).

At the other end, dependency theorists emphasize the exploitive relationship between the developed capitalist metropole and the stagnant, powerless periphery (Foster-Carter, 1985). The emphasis, while purely political in form, is on the structural features of the development process. It seeks to explain the impediments to development not as characteristics of individuals or groups, but as artifacts of structural-organizational arrangements.

Both approaches suffer from a lack of utility at the project level because fundamentally they are ideologies and not theories. To risk oversimplification, modernization theory points to the internal constraints to development such as attitudes and cultural beliefs--in effect, a form of *blaming the victim* (Ryan, 1976). In contrast, dependency theory highlights external constraints such as multinational corporations and foreign investment--in effect a self-defeating mythology of *blaming the bad guys*. While both orientations explain some aspects of the development process, neither offers a technology of organizational design which stimulates local participation and control or project goal attainment.

Development theory has had greater utility at a macro-level of analysis. On the other hand, applied anthropologists have identified local, micro-level impediments to the development/change process. Foster (1973) has listed cultural and social barriers to development which may impede organizational or technological change at the project level. They include traditionalism, fatalism, ethnocentrism, pride and prestige, cultural incompatibility, superstition, group solidarity, public opinion, factionalism, vested interests, local authority, caste, clan, and class. Depending upon the site specific characteristics of the target area, any of these factors may impede efforts to introduce changes in irrigation organization.

Applied anthropologists argue compellingly for the need to adapt development plans, analytic techniques, and organizational structures to the demands of the local socio-cultural environment in order to encourage local participation, enhance project success, and avoid unintended effects. What has been missing is a conceptual framework for designing organizations in irrigation and rural development that stresses the positive opportunities--those general organizational factors which, when incorporated, stimulate efficiency, effectiveness, and goal attainment. It is likely that organizational theories hold unexploited promise for enhancing the achievement of project level objectives in development--particularly with respect to problems of irrigation bureaucracies, farmer organizations, cooperatives, and water users' associations.

Human and Social Ecology Theory

Development theories provide insight into the macrosociological dimensions of the development process. However, these theories have limited utility for enhancing our understanding of mezzoscopic and microscopic aspects of irrigation organization. By contrast, human social ecology theory and organizational theories lead one to a more holistic understanding of irrigation development through their focus on the relationships between individuals (microscopic), the irrigation community (mezzoscopic), and the environment (macroscopic).

The concept of environment has been one of the most important legacies of the Darwinian Revolution. For one, it led to a recognition in most sociological and psychological theories of the importance of external environmental factors in determining the behavior of individuals and groups. The later emergence of human ecology as a theory in sociology marks the recognition of the importance of the interactions between organisms, species, and habitats.

Human or social ecology has provided a shared language between human biology, geography, anthropology, and sociology--a framework for describing human behavior in relation to the other sciences. It also ties in well with the emerging conceptual framework of *sustainable agriculture*. Although to some extent the framework is metaphorical, it avoids the pitfalls of organicism because instead of emphasizing stasis, ecology is dynamic. The key concept in ecology is community--the subset of the species whose reactions to the habitat and coactions with each other constitute an integrated, dynamic, symbiotic system which seeks equilibrium. There are several central ideas in human ecology theory that define the approach and which show clear relevance for irrigation development if we are to describe irrigation systems as communities. At the most fundamental level, communities respond to and are shaped by the environment which also includes the individuals occupying that habitat. Individuals likewise are shaped by the environment of which a key element is the community in which they survive. This allows for the basic insight that behavior is a product of all facets of the environment--the physical, biological, geographic (spatial), and socio-cultural. Correspondingly, the four variables of interest to human ecology theory are: population, organization, environment, and technology. Not coincidentally these are among the four most important concepts in the sociology of irrigation.

By studying human communities, proponents of the view have derived several elementary principles of human ecology. For example, ecologists propose a principle of interdependence arguing that sociality is a given. Interdependence among the species is irreducible and based on symbiotic differences such as the division of labor and commensal similarities such as shared social characteristics. By virtue of these attributes, humans are compelled to both cooperation and competition--the framework for social organization (Hawley, 1973).

Clearly the principles of human ecology could form the basis for a sociology of irrigation. Irrigation is a human activity that is fundamentally rooted in social interdependence and the complex differentiation of functions and roles. Moreover, it is an activity which is inherently fraught with social conflict as users compete for scarce resources and social change, as the irrigation system inevitably responds to fluctuations in the metereological, agricultural, economic, and cultural environments.

The advantages of the human ecology approach are that it links up well with the terminology and theories of agriculture and geography. It provides a handy vocabulary with relevant terms such as niche, population, territory, symbiosis, entropy, etc. It accepts change and conflict as givens and yet is not within an ideological tradition. Among its disadvantages are that it does not handle important social concepts (such as norms) very well. One is left wondering what motivates human behavior in the sense of values, customs, and beliefs. The ability to understand human social organization in the context of an interacting biological population is a very fruitful beginning, but is lacking in insights regarding what we might call the economic and normative bases of society.

Coward's (1980) research provides a contemporary example of the application of human and social ecology theory to irrigation development.

Organizational Theories

The irrigation enterprise involves the effective linking of both technical (hardware) and social (software) elements. The interfacing of the irrigation community with the technical apparatus of irrigation inevitably involves complex organizations (associations/bureaucracies). Given the rich tradition of theories about bureaucracy and organizational behaviors in the social sciences, it is surprising to see so little application of organizational theory to irrigation development. A brief review of some of the major organizational theoretical perspectives will reveal the substantial potential of contemporary organizational theory for enhancing our understanding of the social organizational dynamics of irrigated agriculture.

Many researchers have attempted to trace the historical evolution of organizational theory by categorizing dominant thinkers and writers into periods or perspectives (Harmon & Mayer, 1986; Grusky & Miller, 1981). Nearly all writers on the subject of complex organization begin with the seminal work of the sociologist Max Weber who developed an ideal type of bureaucracy by studying emergent organizational forms in late nineteenth century Germany. For Weber, bureaucracies were goal oriented and driven by rationally derived rules, regulations, and procedures (Weber, 1947):

> A particular goal has been specified and a collection of persons are engaged in a series of separate interrelated and rationally organized activities that presumably will result in goal attainment. The focus of attention is then on the legally prescribed structures and the mechanisms by which they are maintained. Persons comply with organizational rules mainly because the ends achieved by the total structure are valued and each must do his own part if the goal is to be attained (Haas & Drabek, 1973:26).

While Weber's *ideal type* has had considerable value in directing research to key variables in the explanation of organizational life, his work is also replete with shortcomings which provide the stepping stones for refinement and the emergence of alternative perspectives. His work, for example, does not address the importance of *official vs. unofficial* goals, *informal* or spontaneous organization, the relationship of the organization to its environment, and so forth.

Scientific management as espoused by Fredrick Taylor in the late nineteenth century in the United States, shared with Weber, a concern for organizational efficiency (Gross & Etzioni, 1985). Whereas Weber saw efficiency as an outcome of rational organizational rules, Taylor's focus was on ways to motivate workers through economic incentives and by standardizing work procedures into minute components. Time and motion studies provided the basis upon which organizational success could be achieved. In contrast to Weber, Taylor's theory of organizational management does recognize the importance of informal organization and its impact on organizational behavior. A shortcoming of scientific management derives from the view that humans are machine-like in their response to economic rewards (Harmon & Mayer, 1986).

The *human relations perspective* in organizational theory grew out of the Hawthorne experiments conducted by Elton Mayo during the 1920's (Champion, 1975). In the classic experiments conducted in the Cicero, Illinois plant of the Western Electric Company, it was found that workers were motivated by many factors--only one of which was economic. The researchers discovered the importance of status, prestige, appreciation, recognition, accomplishment and other social elements (Roethligberger & Dickson, 1950). From this research it became apparent that informal organization was a paramount determinant of worker performance and attention to group participation led to higher productivity. The participation hypothesis has been effectively applied in the irrigation development setting in a variety of cultural contexts (Bagadion, 1988; Uphoff, 1986a). An important limitation of the human relation perspective was its tendency to treat organizational life as though it existed in a vacuum, ignoring the role of environmental constraints.

The *systems perspective* in organizational theory evolved as a way of visualizing organizations in relationship to their environments. It is important to note that Weber, Taylor and other early writers had a pre-occupation with *formal organizational* structure. By contrast, the systems perspective emphasizes the organization as a holistic unit striving for balance. Additionally the systems orientation views every organization as having a unique *personality* based upon organizational history and the cumulative informal relationships of the group's personnel. The open systems approach emphasizes the interaction (conflict and cooperation) between organizations and their multitude of environments, and interactions between the various subunits that comprise the complex organization (Scott, 1981; Hall, 1991).

Among organizational theories, *the conflict tradition* has proven useful in analyzing irrigation organization. The conflict perspective conceives of change (rather than homeostasis) as the distinguishing feature of groups, because people vary in power (Jesser, 1975). Contemporary conflict theory in organizational sociology has its historical origin in the nineteenth-century writings of Karl Marx. Central tenets of conflict theory in the United States were outlined by C. Wright Mills (1959) and Lewis Coser (1956). The foci of conflict theory are on class conflict (owners vs. workers) and the ubiquity of group contest. Conflict is not confined to social classes but occurs between any groups which compete for the scarce rewards (resources) of society (e.g. racial, ethnic and religious groups, or between farmers and agency bureaucrats). Importantly, conflict does not always imply violence but can be exemplified by disputes and other contests. Conflict theorists tend to be preoccupied with power relations between privileged and dominant groups in competition over scarce resources with subordinate groups.

Karl Witfogel's *Oriental Despotism: A Comparative Study of Total Power* (1957) represents one of the earliest attempts to develop a theoretical framework to study irrigation development. His work draws on the conflict tradition and attempts to explain the despotic character of ancient, large scale agriculture and its relationship to hydraulic society.

In the broadest sense, Witfogel's work represents a theory of societal change. His central argument is that societies tend toward despotism (total coercive power) when the productive base of the society is large-scale, state-managed, irrigated agriculture. Simply put, this is because large-scale systems need a managerial bureaucracy to control various irrigation activities (construction, distribution, flood control) and to mobilize large numbers of the agrarian community to participate in a cooperative ongoing manner in these tasks. The result of the evolutionary process is a coercive managerial bureaucracy which becomes a ruling class with total power over other members of society.

Moving water, especially large volumes, requires sophisticated patterns of organization, technical expertise, and cooperation not found in rainfed agriculture. As Witfogel (1957:18) observes:

Hydroagriculture, farming based on small-scale irrigation, increases the food supply, but it does not involve the patterns of organization and social control that characterize hydraulic agriculture and Oriental despotism....

Thus a number of farmers eager to conquer arid lowlands and plains are forced to invoke the organizational devices which--on the basis of premachine technology--offer the one chance of success: they must work in cooperation with their fellows and subordinate themselves to a directing authority.

Hydraulic economics have special features not found in rainfed agriculture. The division of labor is much more complex, involving preparatory activities (getting water to the fields) and protective activities (flood control) working in tandem. The size and scope of these activities in large-scale irrigation require the coordination and cooperation of many individuals. Coordination was solved by the development of managerial bureaucracies and cooperation is addressed by claims on corviable (forced) labor.

Witfogel's theory of despotism and total power in ancient irrigation systems provides a useful framework for understanding the historical context in which irrigation development has evolved. However, Witfogel's nearly exclusive focus upon power and oppression tends to obscure the stable patterned cooperative social relationships which have been observed in both ancient and contemporary irrigated societies. Conflict theory's preoccupation with power and conflict, while distinguishing it from many other theories, can impede the sociologist's ability to see patterns of consensus, harmony, and cooperation in social relationships. Conflict theory also tends to be concerned with macrosociological issues at the expense of explaining conflict at the community level.

Rational Choice Theory

A tradition of sociological and anthropological writing on farmers in the developing world and the underdeveloped regions of the industrialized world has emphasized the *rationality* of the farmer. Perhaps in response to a reactionary and elitist view of farmers as ignorant, lazy, fatalistic, and incompetent, which is heard far too often in the environs of agriculture ministries and the like, social scientists have noted that the apparently conservative and risk-aversive behavior of farmers is actually a rational response to the fragile economics of peasant society.

Foster (1973) and others have observed that the risks of failure in traditional agriculture have enormous consequences to those who live on the margin of survival. In reviewing studies of peasant agricultural societies, he notes that the observed resistance to change, unwillingness to compete, and static economies are best explained by the *image of limited good* or a zero sum culture in which one person's gain is seen as another person's loss.

The insight of explaining traditional agriculturalists' behavior by assuming that they are rational and are responding to economic forces is the beginning of a different way of looking at irrigated agriculture--one which we would also apply to other important actors in irrigation development such as bureaucrats.

Public choice (and its corollary, rational choice) theory is the application of economics to the study of non-market decision-making. The primary unit of analysis in these perspectives is not the larger society nor the community, but rather the interests of the individual. Yet it is not the individual whom rational choice theorists seek to understand--this is a task for psychology. They instead attempt to understand society and social policy by studying the decision-making of self-interested individuals who seek to maximize their gain and utility through the exercise of rational free choice (Buchanan & Tullock, 1962; Coleman, 1990).

The origins of rational choice theory are in utilitarianism, a social philosophy which, in contrast to Marxism, makes pessimistic assumptions about human behavior. As opposed to Marx, Rousseau, or Locke, the utilitarians argue a realist perspective which would hold that experience rather than reflection reveals human interests and explains their behavior. Humans are empiricists who are driven more by sanctions than by moral abstractions. Furthermore, humans are assumed to be rational economic actors in a utilitarian sense. That is to say that all things being equal, they will choose pleasure over pain and more over less. The assumptions of the *economic person* do not rule out irrational or altruistic behaviors. Instead these types of behavior, excepting mental illness or defect, can be better understood as rational and self interested when analyzed from the individual's point of view. As we shall see, it can be entirely rational and self interested to behave in a manner which the outsider might describe as altruistic or *irrational*--particularly in rural developing communities (Lusk & Parlin, 1986).

A few fundamental ideas can be extracted from the rational/ public choice perspectives to illustrate. Individual choice is the basis of collective action and social organization. What is conceived of as social organization is the structure and systematic interaction of individual choices. Individual decisions are the expression of different preferences and incentives; therefore conflict is inherent in social life and organization is the means of managing that conflict. It follows that rules, norms and discipline are needed to adjudicate conflicting preferences. Finally, conflict produces social change as societies adjust to the dynamics of conflict management.

The key concepts of rational choice which have relevance for understanding social organizations tend to have an economic theme. This is because the theory is a wedding of economics, political science, and sociology as they apply to the explanation of decision-making and social behavior. To illustrate in a very condensed way, participation in collective social action always entails costs to the individual of time and energy. Collective action is more expensive than individual action and is therefore only logically justifiable when its benefits outweigh the costs of non-collective individual actions. Collective actions tend to limit individual liberty so people choose to avoid them when possible. Nonetheless, collective action can achieve results which are clearly impossible through unorganized individual behavior (Lusk & Riley, 1986).

Individuals who do not benefit from collective action will tend to ignore, resist, or boycott such actions unless compelled by force to comply. Therefore, decision-making, to be effective and equitable, must be democratic so that individuals can protect themselves from the actions of the collectivity. Decentralized decision-making incurs fewer individual costs because it is more proximate to the needs of those affected--it is therefore more effective. Centralized decision-making, while less effective, may be more efficient.

Like private market decisions, political or collective decisions are highly important to the users of public goods and services. Therefore the logic of market decision-making can be used with great utility in explaining public actions. Politics cannot be separated from the constituencies which are affected by public actions. Doing so will tend to produce alienation, disengagement, antagonism, and corruption. The task of public management is in large part to link the managers to the constituents through political accountability--democratization. In sum, this means that the relevant public must hire and control the public servant.

In the absence of competitive or marketlike forces being present in the public sector, we can anticipate the emergence of public monopolies or the domination of decision-making by special interests. As in the market, public monopolies can lead to bureaucratic inefficiency, isolation, and corruption. Costs will tend to become magnified and effectiveness and efficiency will both be reduced (Shaw, 1987). The alternative is to transfer as many responsibilities to the private sector as is functionally appropriate and to design public agencies that have incorporated elements of the marketplace. This can be interpreted variously. We contend that public agencies such as irrigation districts or agriculture ministries respond most effectively and efficiently to their constituents' needs when they are multi-levelled, segmental, and decentralized. This allows for the organizational character to be finely tuned to the constituency, the technology, and the policy.

By emphasizing the maximization of individual utility, there is an implicit theoretical emphasis on individual choice, democratic administration, and freedom from coercion. Under these assumptions, organizations are defined as effective if they maximize the individual gain of their constituency while promoting freedom. They are efficient if they produce a net gain for members over aggregate individual actions in relation to social costs.

Two fundamental aspects of organization contribute to effectiveness and efficiency: decentralization, and the incorporation of market forces. Decentralization is important to effective administration because it allows for the adaptation or *fine tuning* of organizational policies to specific, local constituencies. In addition, segmental organizations require a lower investment to mobilize them. Organizations are more competitive (and therefore efficient) if they respond to their specific market--the local constituency. This is made possible when organizations are representative (democratic) and do not enjoy a monopoly (competitive). The prevention of monopoly formation, whether state or private, requires that other agencies be permitted to render an equivalent service and freely compete for clients. Organizations become accountable when democratic representation makes them politically responsive and the market makes them economically competitive.

The rational choice approach thus generates a critique of centralized state bureaucracy--an organizational type which is very common among irrigation organizations in developing countries. The large public bureaucracy inherently tends toward aggregate, unidimensional decision-making rather than multiple, diverse, local decision-making. The interests of the few are sacrificed to the state definition of the collective good--a definition which is not influenced by representative political processes but rather by appointed technicians. The large bureaucracy is neither cost nor decision accountable to its market (constituency) because alternative agencies are not permitted to compete for clients (monopoly), supply and demand are not freely balanced, and organizational costs are not contained for lack of incentives to do so. Also the coercive powers available to the state to force compliance to policy divert power from free constituents to unelected technocrats with the effect that the organizations do not have incentives to serve their relevant public (Ostrom, 1974). Given the lack of accountability, decisionmakers are more subject to the corruptability growing out of interest-group control and the abuse of power.

The alternative is to limit and control administrative power; to stimulate competition through decentralized, multi-organizational arrangements; to maximize efficiency by reducing expenditures of time, effort, and resources; and to incorporate representation from relevant consumer constituencies--in sum, democratic administration (Ostrom, 1974).

Rational Choice and Natural Resources

Rational choice is particularly useful in the study of natural resource and water management problems because: (1) it provides a theoretical basis for fitting organizations to resource type, (2) it emphasizes the analysis of incentives in resource use and, (3) it readily applies to common resource management problems such as resource depletion, negative externalities, *free riders*, and monopolies.

A basic social science question in natural resource and irrigation water management is the determination of the most appropriate organization or institution for managing resource goods efficiently. The most efficient, effective and, therefore, appropriate institution, is a function of the nature of the good--a question of fitting organizations to the type of resource being managed.

Three fundamental types of goods can be identified: public goods, private goods, and common pools. Their differences are best understood in relation to their exclusivity, divisibility, and subtractibility.

Public goods are nonexclusive in that they are equally available for consumption to all of the members of a population and nonsubtractible because one individual's consumption of the good does not subtract from the amount available for another consumer's use. National defense, air, and public broadcasting are examples of public goods.

Private goods are exclusive, divisible, and subtractible. Marketable commodities and real estate are examples of private goods--they can be broken up into units (divisibility), excluded from multiple use (exclusivity), and one individual's use of the good reduces availability to others (subtractibility) (Goetze, 1986).

Common pools are subtractible, nonexclusive and not easily divisible. Therefore, they combine characteristics of both private and public goods. Examples of common pools include public rangeland, fisheries, and lakes. The nonexclusive character of common pools can lead to a dilemma of overuse and depletion because the resource is subtractible, but is held in common by a community of users, all of whom have access to it. The logic of the unregulated commons is that individuals will draw on the resource to maximize private benefits and pass on the contingent use costs to the collectivity. In the absence of institutional restraints to overuse, the resource may be exhausted as increasing numbers of self interested users consume the good (Bullock & Baden, 1977). This logic is useful in accounting for overgrazing, deforestation, pollution, water mismangement and irrigation deviance.

To optimize organizational efficiency, the nature of good can be matched to institutional type. Logically, public goods correspond to state responsibility and private goods to management within the free market. A nonexclusive, nonsubtractible good (public) can be allocated in the collective interest through representative government. Problems which may arise in this process include state-imposed inequities in resource access, the differential application of norms based on individual status and power, noneconomic subsidies for projects which could be self-sustaining, state monopoly by a subset of users, bureaucratic insensitivity to user concerns, corruption, patronage, nepotism, bureaucratic passivity, and related problems of non-representative management (Lusk & Riley, 1986).

An exclusive, subtractible good (private) is efficiently allocated in a marketplace where user prices correspond to costs and demand. Problems which can arise in this respect include the diffusion of negative externalities such as pollution, social inequities in allocation, and private monopoly control (Sproule-Jones, 1982).

With respect to common pools, some have argued on behalf of public monopoly control (Baden, 1977), while others have suggested that the marketplace can most efficiently allocate the good (Ostrom & Ostrom, 1975). In either case the corresponding problems of management noted above may emerge. A more useful analysis points to the divisibility and multiple ownership aspects of a common pool to identify an appropriate institution (Goetze, 1986). If a common pool can be unitized (divided) into portions and distributed to multiple individuals based upon their willingness to pay, it may be more efficiently managed as a private good. Surface irrigation water meets these criteria. If, on the other hand, the commonly held good has integrity only as a single unit and is not transportable, it corresponds that representative collective ownership (state or private) is the most appropriate institutional type. Instream fisheries meet these criteria.

The challenge of common pool management is to prevent the dilemma or *tragedy of the commons* wherein individuals perceive that their marginal use of the resource in the short term is inconsequential to the final outcome. The result is that multiple users will eventually deplete or degrade the common good. The dilemma of the commons reflects how the interests of resource users can come into direct conflict with public welfare when individual incentives do not correspond to policy objectives. Even with renewable resources this depletion of the good may surpass the point of no return. An aquifer or pasture can be exhausted to a level from which it will not regenerate in the foreseeable future (Veeman, 1978).

It is crucial that appropriate institutions be selected to manage a resource so as to prevent the problems which can result from a mis-match between the organization and the type of good (Goetze, 1986). In China, for example, considerable effort has been expended in re-directing the management of agricultural resources away from collectivist strategies. While this has been effective in increasing domestic food supplies and fostering competition, numerous examples can be identified of situations in which public or collectively held resources have been depleted in the name of privatization or what the Chinese called the *responsibility system*. Schell (1984;77) has reported that as the Party retreated from its role in managing public goods such as dams, terrace walls, flood control and irrigation projects, many of these structures have fallen into disrepair, and pumps, concrete blocks, wiring, and motors have been stolen or sold as scrap. Government *ideological work* to appeal to the responsibilities of the systems' users will prove insufficient in the absence of collective organizations which enforce management policies over goods that have fallen into the common pool.

The technology used to exploit a resource can also be fitted to organizational type. The two relevant dimensions of this analysis are *scale* and *divisibility*. Freeman and Lowdermilk (1981) have argued that a divisible technology, one that can be used in small scale, portable units (seed, fertilizer, handtools) requires a much lower level of organizational investment for utilization. The market can optimize the allocation of divisible technology. In contrast, a major irrigation project involves large-scale *lumpy* technologies such as dams or lined canals which are not divisible or portable and which require high organizational and capital costs for implementation and utilization--a role suited to the public sector.

State management is usually fitted to large-scale technologies that serve multiple constituencies because of the representative and mediative (judicial) functions. This is particularly true when dealing with *rights of way*, equity issues, minority group rights, and taxation. When the decision costs for resource management (the time and energy invested in securing agreements among and between constituents and interest groups) are very high, as with very large or diverse user organizations, a state role can be a useful option if those who must manage, act on behalf of their appropriate constituency. Private stockholder organizational forms, such as the corporation, can be very effective in coordinating large-scale technologies for singular constituencies. Note, for example, that the majority of dams and reservoirs in the United States are privately owned and operated. If the technology is of such a scale that the state's taxation power must be invoked, rights of way adjudicated, or if the technology benefits assorted political or social constituencies, state management is implied.

Without careful attention to the organization of incentives, there is often a direct conflict between individual interests and collective welfare in managing both common pools and public goods. This suggests that great care be exercised in designing organizations so that there is a high degree of correspondence between individual payoffs (benefits) and public policy. Seen in this light, efforts to appeal to the altruism of resource users or to *bureaucratically re-orient* or sensitize resource managers seem naive (Korten, 1980). Such efforts could be better invested in designing organizations that efficiently manage resources in the public interest by retaining incentives for individual use. This process can be abetted through privatization, democratization and decentralization.

Rational Choice and Irrigation

Irrigation organization and rural development require collective decision-making as farmers, bureaucrats, and other interested parties express their political will by attempting to manage the water resource in their own best interest. The structure of this decision-making process is determined by the legal, political, economic, and cultural environment. Of particular interest to the success of irrigation development is the local political economy of agriculture. A rational choice analysis of irrigation organization will, therefore, emphasize: (1) the nature of the good, (2) the organizational character, and (3) the incentive structure. The purpose of the analysis is to arrive at a site specific maximization of *appropriate use* which can be defined as the organizational design which promotes efficiency, equity, and project goal attainment.

Privatization and the Nature of the Good

Irrigation water is a private good. It is divisible (can be readily unitized), subtractible (one irrigator's use subtracts from the total available for others), and exclusive (boundaries between uses can be maintained). *If private rights and responsibilities over irrigation water are not established, however, it becomes a common pool.* In order to achieve policy objectives such as equity, some governments may define irrigation water as a public good. Yet unless everyone has equal and non-subtractible access to the resource, it cannot truly be a public good. To legally define irrigation water as a public commodity is to formalize its common pool character with the consequential risk of depletion and maldistribution.

Even though irrigation water is a private good, the property rights can be assigned to the public sector. In this case, its use is determined by the pressures of interest groups, elites, or legislatures on government agencies and bureaucrats (Goetze, 1986). State control carries with it a high risk of monopolies by a subset of users, non-accountability to users, inefficient use, and corruption. Alternatively, property rights can be assigned to the market where the cost of water will correspond to its productive utility for individual or multiple users. While this will tend to increase efficient use, private sector risks include inequitable distribution, monopoly control, and negative externalities such as pollution, salinization, and soil erosion--risks which can be mitigated through careful organizational design.

Water Shares

The nature of the good is fundamentally linked to the institutional alternatives for its management. This is, of course, based on the central importance of ownership and tenure to economic efficiency. At the simplest level, the question is whether state or private ownership is best suited to the resource type. If irrigation water is privately claimed, the role of the state is to adjudicate rights and to arbitrate in disputes. If irrigation is of the common pool character, either because it has not been made exclusive by government or been claimed by private users, it can be unitized (or divided up) through share systems. The alternative is a dilemma of the commons. Finally, if irrigation water is strictly state-owned, there is no capture of economic rents by private actors nor any market incentive to control costs, increase efficiencies, conserve the good, or maximize production. The case for establishing property rights or otherwise privatizing irrigation can be compelling when it is balanced with state involvement in controlling negative externalities, protecting minority rights, managing disputes, and capitalizing large projects benefitting multiple constituencies (Coward, 1986a).

Broadly defined, every irrigation system involving multiple farms and limited water is based on private shares. These shares may be explicitly and formally identified as legal rights and responsibilities (common in mature irrigation schemes) or may be informal and consensual. Because a share system is in place, however, in no way guarantees that it is equitable, fair, or productive. Indeed share systems generally mirror the distribution of rights and benefits in the broader social order: democratic, oligarchical, egalitarian, pluralist, theocratic, statist, etc. As a result, many irrigation systems are plagued with *tailender* and *free rider* problems or other inequities which produce social conflict, irrigation inefficiencies, and poor production. The key social and legal mechanism for organizing and understanding irrigation water management is share systems. A share system determines the property rights of water users by defining the volume, timing, and contingencies of water allocation and delivery.

The design of a share system is, in effect, the social engineering of irrigation. Shares are the social and legal basis for the organization of water under a given irrigation technology. To illustrate, a share system can be designed for local circumstances using one of the above types of property rights or some combination thereof. For example, in the Spanish *Acequia* system of Northern New Mexico, USA, many of the *acequias* (ditch associations) use a mixture of the priority by farm characteristic and priority by crop. In their situation they have chosen to recognize the primacy of early settlers' rights by conferring shares on the basis of *first in time, first in right*--a method commonly used by state governments throughout the American West. Interestingly, though, the acequias also recognize the importance of subsistence food crops to rural welfare, and so during dry seasons or periods of drought, family food gardens are given primacy in rights over cash crops. Two additional considerations are built into the share rules. The water must be put to *beneficial use* as defined by state law--water cannot be wasted or the right to use it is lost. Additionally, acequia officials, in designing their rotation schedules, must take practical considerations such as ditch losses and the location of a field on the system into account so that irrigators get a just and proportionate share (Lovato, 1974). In this sense, the New Mexican approach is to have all of the farmers share equally in the conveyance and evaporation losses along the system. There is no maldistribution between the head and tail.

It can be argued that problems of maldistribution are inherent in gravity-fed surface irrigation, but there is no reason to assert that the engineering of irrigation necessarily determines the rules of allocation. The logic of shares is influenced by the engineering environment, but not determined by it. Tailenders do not get less water because of seepage and conveyance losses, but rather because the farmers on the canal are not sharing the losses or *shrink* (see Chapter 2). Lining canals may reduce seepage but it will not necessarily remove the inequities of a tail problem.

In some share arrangements the rotation actually reinforces inequitable distribution. The *warabundi* share model of Pakistan combines shares by time period and farm size with priority by location in such a way as to give the tail farmers proportionately less water because they must absorb the conveyance inefficiencies. Thus a central question in looking at the efficiency, productivity, and equity of an irrigation organizational design is: Does the share system promote or diminish the problem of head and tail? (Freeman, 1986).

While a public choice orientation will generally argue in favor of greater privatization of water rights than is typical in order to improve use efficiencies, cases can be found where multiple private ownership has been relinquished to state regulation so as to prevent a common pool dilemma. In the West Basin Aquifer of Los Angeles County, California, USA, joint users of underground water functioned under a Doctrine of Absolute Rights wherein their water rights were tied to ownership of land above the aquifer. As the water demand grew, the underground basin was being depleted beyond safe yield levels (the future viability of the aquifer was endangered due to saltwater intrusion). To prevent a depletion which would negatively affect all of the users, a Doctrine of Correlative Rights was implemented through state courts and the creation of a public water district. User rights were adjudicated based on safe yield levels and the pattern of historical use (Blomquist & Ostrom, 1985).

The West Basin water users autonomously developed the institutional capacity to manage a commons in the collective interest. The case illustrates that while resource users will act in their own rational self interest, collective and/or state-monitored management can be an effective organizational choice when private share rights are retained and representative management is ensured. It would be more economical over the long term to collectively regulate use than to follow the individual pumper's incentives to increase unrestricted use to the point of depletion. What is notable about the West Basin case is that the users self-imposed new cost sharing and institutional controls on water use without state coercion. It can be argued that the correlative rights doctrine could be of significant value in organizing groundwater rights in the Indo-Gangetic Plains of Pakistan and Northern India where presently groundwater pumping is essentially an unrestricted common pool (Veeman, 1978).

User Fees

The privatization of irrigation is also related to costs, productivity, and farmer participation. It is dysfunctional not to price water in the marketplace. To make water freely available to users without imposing the corresponding costs of diversion and storage is to create a common pool and to transfer the operation, construction, and maintenance costs to the state (in effect to the taxpayer who may not be a project beneficiary) or to foreign donors. While in traditional subsistence economies, it may be unrealistic to expect risk-aversive peasant farmers to bear the front end costs of major project development, it is not unrealistic, indeed it is desirable, to have them bear the costs of ongoing operation and maintenance if they also have representation in determining the costs, rules of allocation, and methods of resolving disputes.

Free or very low-cost water encourages overuse, reduces the incentive to cooperate and participate in irrigation organizations, lowers system productivity due to overapplication (overirrigation can reduce yield because of inadequate root aeration) and poor conservation practices. In a very low cost state-subsidized or gratis situation, there is no incentive to husband water. If water is not abundant, overall system productivity will be reduced because of uneven application across the scheme--headenders will tend to overirrigate and tailenders will experience reduced yields for lack of an adequate supply. Participation will be minimal because the organizational costs of increased collective management will exceed the existing costs of unrestricted use (see Figure 1.1).

Very high cost cases encourage conservation but at the expense of system productivity. Unless low profit dryland crops are used with supplemental irrigation, the alternative is lowered yields due to crop stress from underirrigation. Artificially high priced water raises input costs to the level of non-profitability. Farmers have an incentive to disengage from irrigation organizations that impose unreasonable costs, so participation is minimized.

If the market is allowed to freely function, an optimization point can be achieved at which water costs correspond to demand and use, and system productivity is maximum because the incentive to conserve is balanced against the need to achieve optimum crop application efficiency (see Figure 1.1). In this case, farmer participation is high because of the individual incentives to keep the organization responsive to the local farm economy. For the model to work, it is crucial that the irrigation organization be democratic and decentralized.

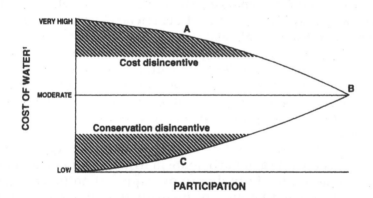

A. High cost water, low productivity, low participation
B. Moderate cost water, high productivity, optimum participation
C. Low cost water, moderate productivity, low participation

'Water costs are both organizational and per unit

FIGURE 1.1 Farmer Participation and the Cost of Water.

Democratization and Organizational Character

By virtue of having to share the natural resource, water users are inextricably connected by the physical distribution system and through the socio-political organization managing that distribution. The control of water availability by users is a function of the technology type and the organization's scale and character.

The simple case of gravity flow, canal-fed, surface irrigation can illustrate. On-farm availability in this situation is a function of the number of upstream irrigators on the delivery system, total water supply, net seepage, and evaporation. To have any effect on the delivery of water to the field channel, the agriculturalist has options that correspond to the technology. The farmer can try to increase total water supply to the scheme, reduce seepage through canal lining or other techniques, or can work with other farmers to collectively address inequities in downstream delivery. All of these strategies represent organizational problems. To affect total water supply requires that the farmer reach up into the system to influence those who control the main works. To line canals or otherwise reduce water losses necessitates influencing those who control the middle level irrigation organization: the canal company, water users' association, etc. To address inequities in the delivery of water to the tail requires that all of the users on the system cooperate in an allocation method that is fair and proportional. We can see that each method of positively affecting on-farm water availability requires a high level of farmer organization because individually a user cannot effectuate significant changes in water management except at the field level.

The function of irrigation organization is to design and manage the institutions and physical structures which economically deliver water in a timely and reliable manner with the highest possible degree of control at the farm level. Water has little or no value if it is not of sufficient volume for crop needs or arrives at the field channel too early or too late. The volume of water must be predictable throughout the growing season so that growers can plant in relation to anticipated supply and the predicted volume must be available when needed and scheduled.

Fundamental to the effectiveness and efficiency of irrigation organization is the problem of accountability and control. Because of the need for timely applications of requisite volumes at the field level in order to maximize yield, irrigation organizations that are not controlled or at least influenced by the irrigators themselves will produce inefficiencies (Parlin et al., 1986). This principle of irrigation organization functions because of accountability. The actors in inefficient irrigation schemes usually do not have to bear the costs of their inefficiencies. If, on the other hand, those who bear the organizational costs also capture the benefits, we can expect such inefficiencies to decline. To build in accountability is to ensure that those who must take the risks and pay the costs of farming should also be able to capture the benefits. Field experience reveals that farmers are willing to make enormous investments of energy, labor, and cash when they perceive that they are in control, their risks of failure are low or moderate, and that they will be able to reap the benefits of their work. What we have often seen in the field, however, is that the primary users of water have little or no control over its administration or delivery. Individuals whose livelihood is not dependent on the efficient and timely delivery of water (bureaucrats) are typically those who have the greatest say in how it is allocated and managed.

The argument in favor of the bureaucratic administration of water is based on the presumed need for expert specialists to manage complex engineering and allocation systems for multiple users. This does not obviate the logic of farmer control. It is possible to privatize, decentralize, and democratize the administration of water while still employing technicians and *experts*. Farmer owned and operated waterworks can be managed by elected farmer directors under corporate models of organization. In cultural contexts where the private ownership or management of natural resources is legally restricted, the appropriate public organization can be democratized by electing water administrators or commissioners who then supervise specialists in the interest of the user constituency. Accountability is built in through the participatory process.

Democratic administration is a goal of irrigation organization because of the corruptability of decision makers and the abuse of authority possible under centralized bureaucracy. In resource and other public management, administrative rules are not a matter of political indifference to users (Ostrom, 1974). Indeed farmer welfare is fundamentally linked to the decision-making process growing out of those rules.

The democratization of irrigation organization can be stimulated by a reduction in the scale of such associations. Large groups do not induce a sense of accountability or of permanency. In small groups individuals tend to feel more visible and, hence, more accountable to one another and a sense of reciprocity emerges among group members. In marketplaces where individuals expect to have continued interaction over long time frames, a norm of reciprocity is likely to develop in which individuals recognize the need to cooperate to achieve mutually rewarding pro-social outcomes (Axelrod, 1984; Coleman, 1990). Reductions in scale can also stimulate healthy competition and thus efficiency. Multiple, diverse, segmental organizational forms allow free movement from one association or organization to another as consumers seek to find the least costly organization. Reductions in organizational scale can prevent monopoly (the antithesis of free choice) by permitting competition, change, and face-to-face reciprocity based on trust.

Our working assumption is that farmers rationally seek to control their resources in order to maximize agricultural production and can effectively do so when they have private land and water rights, open markets, and predictable and accountable organizations for resource management. Integrated irrigation organization development pursues these ends.

Three central concepts can be used to guide organizational design for public development: *democratization, decentralization,* and *privatization.* Democratization is the process of building political accountability into organizational design. To decentralize organizations is to break decision-making out of the top-heavy hierarchical mode by transferring authority (and responsibility) to those who are in communication with the needs of the specific local constituencies affected. Privatization is the process of restoring some public functions to the marketplace either by deregulation or the establishment of property rights for what had been publicly owned goods (Lusk & Parlin, 1986).

In irrigation, the democratization of the water management authority or other irrigation organization provides for accountability to the users and funders. Decentralization reduces the machine-like character of bureaucracies by scaling the decision-making process to the corresponding constituency, level of technology, and local environment. Privatization can clarify rights of use and ownership, stimulate competition, and diminish state coercion.

There are numerous implications of this theoretical approach to irrigation development and management. If one accepts the general framework of the theory, the following strategies are suggested: (1) clarification and adjudication of water property rights and entitlements, (2) the formation of private irrigation companies or ditch groups, (3) water marketing, (4) user fees, (5) elected water management officials (from the commissioner to ditchrider level), (6) decentralized segmental irrigation project management, and (7) codification and enforcement of the rights and responsibilities of water users and bureaucrats.

Another key implication of a rational choice model for irrigation development and management is that policy, planning, and organizational design must be cognizant of how the system is viewed from the point of view of the individuals involved in it--the farmers, policymakers, funders, and bureaucrats. When viewed from the farmer level we may be surprised at the confusion of incentives, sanctions, and cultural preferences which shape decision-making. Equally important is how the system is seen by the various bureaucrats. Government officials can also be assumed to be rational and self-interested decisionmakers who act in response to their own set of incentives and perspectives. Their behavior, which is not necessarily in the *public interest*, is no less important to the success or failure of an irrigation project than that of the presumed primary beneficiaries.

Rationality: Bureaucratic and Farmer Participation

Considerable attention has been given by social scientists to the problem of involving farmers and other water users in the process of managing and developing irrigation projects (Uphoff et al., 1985; Parlin & Lusk, 1988). The lessons of this line of research are that farmer involvement in planning, design, water allocation, and conflict management has several positive effects on project outcomes. Studies in the Philippines and Sri Lanka, for instance, have demonstrated reductions in conflict and deviance in addition to improved water application efficiencies (Bagadion, 1985; Uphoff, 1986b). These are findings that are entirely consistent with the broader research traditions of the sociology of organizations and rational choice theory which would suggest that worker or farmer satisfaction and productivity will be linked to the degree to which they as constituents are meaningfully involved in the decision-making process (Blumberg, 1969).

Repeatedly, however, irrigation development specialists report that one of the most serious obstacles to project success is not only the meaningful involvement of farmers, who after all are direct beneficiaries of increased water supply, but the bureaucrats who have little or no incentive to implement policies which have no bearing on their own welfare (Wade, 1982a; Freeman, 1986).

A preoccupation with farmer participation may have obscured to a degree the fact that farmer behavior is partly a function of the organizational behavior of project and agency bureaucrats who interact directly or indirectly with farmers, implement or neglect project policy, and otherwise have a bearing on the outcomes of the irrigation enterprise.

Several researchers have recognized the importance of the interface between the farmers and the bureaucrats. Bryant and White (1984:9), for example, propose, "that if participation is to occur and be effectively managed, there must be incentives for farmers and peasants to participate. There must also be incentives for field level administrators to facilitate that participation."

They and others (Bromley, 1982; Freeman, 1988) have emphasized the importance of farmer-bureaucrat linkages and institutional reform to increased farmer participation.

Korten (1980) and Bagadion (1988) have suggested *bureaucratic re-orientation* as a method of improving the relationship between irrigators and bureaucrats. The *re-orientation* or training technique used is to appeal to the altruism, commitment, or *public interest* of water management officials--an approach that to be effective will also have to incorporate *incentives* (Bryant & White, 1984) and *sanctions* that are built into the organizational design (Lusk & Parlin, 1986).

Increased farmer participation alone is not necessarily a panacea for irrigation project success. Indeed some of the most successful systems in the American West are characterized by an almost complete absence of farmer or user participation. One can attend irrigation company meetings in Utah and Colorado, for instance, in which the users themselves are not participating. This is not because the irrigation company is a failure but precisely because it is a success. The users, having few complaints or conflicts, have no incentive to become involved. The timely arrival of adequate volumes of water to their land has made participation moot. What is important is that the institutional mechanisms for user participation or even better, control, be present in the irrigation organization's design. The irrigation association must be *engineered* or designed in such a way that decisionmakers are compelled to implement policy and represent the interests of the users.

The management of irrigation development behavior involves the design and monitoring of irrigation organizations that can simultaneously implement water management policy and represent the needs of the user constituency. It consists of the coordination of both *push* and *pull* factors which direct the irrigation project toward its stated development objectives (see Figure 1.2). If we conceive of irrigation projects as dynamic organizational systems made up of rational utilitarians, then we can identify those factors or pre-existing conditions that impel or push the key actors to behave according to development objectives as well as those factors which incentivate or pull the individual toward the desired outcomes. To look at an irrigation project in this way is to focus on the policies, rules, sanctions, preconditions, and incentives which shape the behavior of the two most important groups involved in irrigation development: farmers and bureaucrats.

As in any systems-type model, we assume that an irrigation project is dynamic and evolving as it seeks an equilibrium resulting from the forces impinging upon it. Those forces can be consistent with the goals of the project or may mitigate against them. The path of the project's evolution is a function of the development process (one of induced social and technical change) working against the counterdevelopment forces present within the irrigation system and the external environment. The task of irrigation management is to imbalance the equation in the direction of development by juxtaposing inputs and incentives in such a way as to change the behavior of farmers and bureaucrats toward desired project outputs. In addition, attention must be given to overcoming or at least minimizing counterdevelopment forces and incorporating a feedback mechanism whereby the project can become self regulating.

Figure 1.2 suggests that the management of irrigation development must simultaneously address the behavior of bureaucrats as well as farmers. The development inputs and incentives shaping the bureaucratic and farming systems can be coordinated to maximum effect. Note that the profits, status, and search for security motivating farmers have corollaries in the bureaucracy--merit pay, promotion, perks. Similarly, such incentives become more powerful in the presence of those pre-existing conditions and inputs which will make pro-development behavior possible--order, rules, and training.

By using the rational choice model to load the development equation on behalf of success, we can transcend a problem solving or *clinical* framework which would seek to resolve project difficulties on an atomistic, post hoc basis using specialists to *diagnose* project ills. A problem such as farmer non-payment of water fees cannot be seen in isolation from the organizational framework that produces such deviance. Likewise, bureaucratic corruption and patronage can be better understood by looking at the checks and balances that impinge upon bureaucrats. To solve a specific problem will usually require some tinkering with the whole system.

Furthermore, a systems approach that focuses on the incentives and sanctions that motivate key actors will give greater emphasis to project performance and opportunities as opposed to project problems and diagnoses. Therefore the perspective will look for *what is right* with an irrigation system with the goal of building upon what Keller (1988) has called *the islands of excellence.* Superficially a project may appear to be in a chaotic state when the focus is on overgrown ditches, siltation, breached canals, poorly drained fields, and damaged structures. Yet few irrigation projects are immune to such problems and a *technical fix* orientation may overlook positive patterns of cooperation, water sharing, maintenance, and profitability. These issues can be addressed fully only by looking beneath the symptoms to the underlying social structure.

Institutional Reform

We have seen that farmers and workers respond to meaningful involvement in agricultural organizations. This is in part because the project will be more likely to be accountable to their constituency's needs--a group who must survive by selling products in a marketplace that tends to contain their costs and act as an incentive for their efficiency. We contend that there are similar mechanisms by which bureaucrats can become more reliably efficient in attaining project goals. Bureaucrats, after all, spend other people's money, and have no incentives to reduce the size of their budgets. In most situations they are not elected by those they serve and therefore have little reason to be representative of their interests. Rarely are they subject to the forces of competition nor is their personal welfare linked directly to the successful implementation of policy (Knott & Miller, 1987). While it has become clear that irrigation development can be greatly accelerated by incorporating farmer participation into the project, the importance of the behavior of those who are external to the farming system must be considered.

Summary

The application of a rational choice perspective to irrigation development suggests that three organizing concepts be used to design irrigation organizations: democratization, decentralization, and privatization. These ideas can be used with effect not only in thinking about water user associations, but can also be applied to the larger institutional framework of system management.

The strategy implied by this approach to institutional reform will specifically suggest the: (1) implementation of civil service rules and sanctions which are promptly and equitably enforced, (2) replacement of the culture of bureaucracy with the rules of

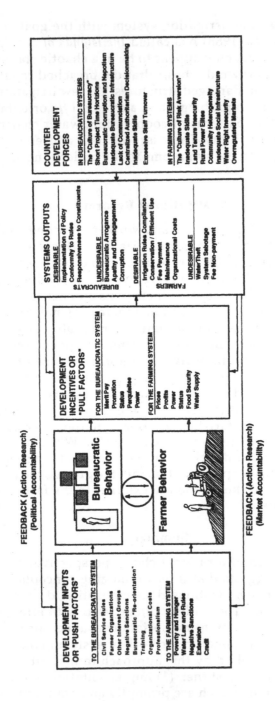

FIGURE 1.2 Managing Irrigation Development Behavior.

meritocracy, (3) design of *representative bureaucracy* built around the election of water commissioners, directors, and ditchriders, (4) development of corporate or utility water management organizations in which users are *stockholders*, (5) clarification and adjudication of water property rights and entitlements, (6) emergence of decentralized segmental irrigation project management, (7) codification and enforcement of the rights and responsibilties of water users and officials, and (8) optimization of use through water marketing and user fees.

While it is clear that such ideas must be carefully fitted to the local social and cultural environment by selecting incentives and sanctions which are appropriate in a given context, they provide a useful starting point for guiding the institutional change which is an inevitable part of any development project.

The application of sociology to irrigation development and management produces its best yields when guided by relevant theoretical perspectives. The eclectic use of organization theory, human ecology, conflict theory and rational choice theory can be helpful to efforts to effectively manage irrigation development behaviors and processes. The use of a systems model which incorporates development inputs, incentives, counterdevelopment forces, environmental factors, and feedback can provide a holistic context for irrigation project management.

2

Designing the Organizational Interface Between Users and the Agencies

David M. Freeman*

The best structure will not guarantee results and performance, but the wrong structure is a guarantee of non-performance. All it produces is friction and frustration. (Drucker, 1974:519)

Creating and operating organizations has always been a central business of human beings who, early on, recognized in all cultures, that they must make permanent arrangements to secure and manage collectively what they could not obtain individually. In large-scale gravity flow irrigation systems, one item that people have never been able to obtain individually is a unit of irrigation water control. Irrigated agriculture, in large-scale gravity flow canal systems, has always meant the organized collective attempt to control water to better fulfill crop consumptive needs. The social construction of rules for employing physical tools to collectively control irrigation water is as old as recorded history. The progress of peoples, in general, has always centrally depended upon how they have organized their collective lives, and the progress of irrigation systems now, as always, depends upon the quality of their irrigation organizations.

Only about eighteen percent of the world's cultivated land is irrigated, but it produces roughly one third of the planet's hu-

*The material published in this chapter is excerpted from Freeman, D., with Bhandarkar, V., Shinn, E., Wilkens-Wells, J., and P. Wilkens-Wells (1989). Local organizations for social development: Concepts and cases of irrigation organization. Boulder CO: Westview.

man food supply (Rangeley, 1987). However, the fact that many landscapes of the world are now dominated by dams, reservoirs, and canals cannot hide a disquieting fact. Many irrigation projects, representing many nations and cultures, have not served the needs of farmers and agricultural production as planners had hoped.

The story of the typical irrigation project in nation after nation, culture after culture, is one of failure to fulfill projected economic returns to investment, and of farmers who not only fail to exploit their relatively expensive water supplies to the degree which had been planned, but who frequently exhibit irrigation behavior viewed by main system managers as detrimental to the functioning of the systems. Montague Yudelman (1987), reflecting on World Bank experience, has suggested that Bank irrigation projects seldom have met expectations. Expressions of disappointment have been many (Bottrall, 1978; Bottrall, 1981a; Bottrall, 1981b; Chakravarty & Das, 1982; Levine, Capener, & Gore, 1972; Lowdermilk, Early, & Freeman, 1978; Pant & Verna, 1983; Posz, Raj, & Peterson, 1981; Reidinger, 1974; Sharma, 1980; Steinberg, 1984; White, 1984). The picture of poor irrigation water management unfolds around low levels of water use efficiency, marked by inequities in distribution, disappointing cropping intensities and yields, and irrigation bureaucracies that perform with insufficient regard to the needs of farmers to control water for production of food and fiber.

The thesis of this chapter is that strategic irrigation project problems are partly a function of the failure to properly integrate main and farm irrigation systems with viable middle level irrigation organizations operating in the interface zone as identified in Figure 2.1.

It is the objective of this chapter to: (1) present an analysis of the problem of irrigation organization breakdown in the middle level between main system central bureaucracies and farmers, and (2) to formulate strategic variables and relationships which can contribute to improved design of such local irrigation organization.

Organizing for Water Control

Reconciling Main System Supply With Farmer Demand

Water control by farmers, the capacity to apply the proper quantity and quality of water at the optimum time to the crop root zone to meet crop consumptive needs and soil leaching requirements,

37

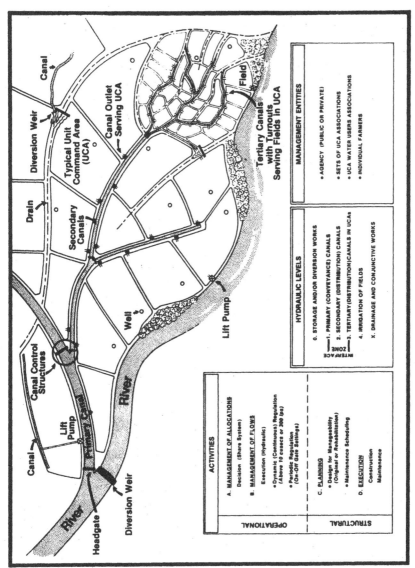

FIGURE 2.1. Irrigation System Management.

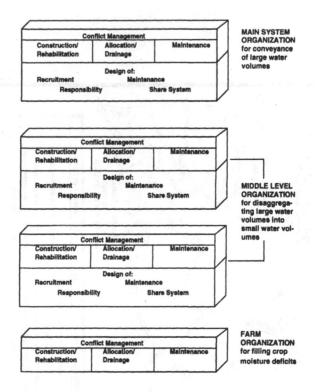

FIGURE 2.2 Levels of Irrigation Organization.

is a fundamental yardstick used to measure effectiveness of irrigation systems. The very idea of water management presumes the idea of control over water supply. Such control, in turn, is a function of the manner in which people become organized at several levels--the main system, and one or more tiers of middle level organization--to provide an interface between main system management and individual water users (Figure 2.2). Water control for main system management in any nation or culture area, must mean something different than water control at the farm level. The meaning of water control shifts as one follows water from main system works to farm application. This shift in meaning necessitates effective middle level irrigation organizations to provide an interface between the very different, even incompatible, requirements of main and farm systems.

Water control is critical, not only to improve production in any given season, but also to sustain the production environment across seasons. Higher water control permits less water to be used per unit of production which translates into reduced energy consumption, soil erosion, waterlogging and salinity (Mathur, 1984; Reddy, 1986). Because high yielding plant varieties are much more demanding of adequate, timely water applications, farmers with inadequate water control will refrain from investment in such varieties and the associated costly inputs of fertilizers and pesticides. As control over water diminishes, it becomes necessary to apply increasing quantities of water whenever available to attempt to assure survival of at least a portion of the plant population. Erosion and associated waterlogging and salinity problems are thereby exacerbated.

Irrigation water management in large-scale gravity flow systems is the process by which bureaucracies capture and control water in central irrigation works and pass it on to local command areas, which divide and control it further. In turn, local unit command organizations (Figure 2.1) pass the water on the farmers who must place it in crop root zones at times and in amounts which make it most productive and least damaging to the production environment.

Years of careful experimentation have established that applying the right amount of water to crops at the right time, as defined by properties of the plant, soils, and climate, is critical to crop productivity. Doorenbos and Kassam have plainly stated the problem in its technical aspects (1979):

> The upper limit of crop production is set by the climatic conditions and the genetic potential of the crop. The extent to which this limit can be reached will always depend on how finely the engineering aspects of water supply are in tune with the biological needs for water in crop production. Therefore, efficient use of water in crop production can only be attained when the planning, design and operation of the water supply and distribution system is geared toward meeting in quantity and time...the crop water needs required for optimum growth and high yields.

The extent to which the water supply can be tuned to crop biological requirements is a function of the organizational operations conducted at three or more levels (Figure 2.2).

At the farm level, control over water is fundamentally determined by the operation of organizational networks established to operate upstream physical structures. The effectiveness by which irrigation water is used productively is a function of the organizations created to construct, rehabilitate, operate, and maintain works, and to resolve conflict. Field control by the farmer is critical. Only the farmer combines the factors of production in a particular field and either succeeds or fails to bring in a crop. If water comes too soon, too late, in amounts too great or too small, the productivity of that water is sharply reduced. Because different plants exert very different consumptive demand under varying stages of growth, soil, and climatic conditions, irrigation water can fulfill that changing consumptive demand only if it has been subjected to rather precise control which allows farmers to be rapidly adaptive in managing it.

A rice cultivator in southeast Asia working in an irrigation system designed to deliver continuous and simultaneous water supplies to hundreds of farmers in a given command, faces very different water control problems to permit tilling, transplanting, fertilizing, and harvesting than does a farmer in Northern India or Pakistan who works within a rotational delivery system to serve the consumptive requirements of wheat or cotton. Yet even within a given irrigation system, the consumptive demand of crops can be expected to be highly varied. The farmer growing shallowly rooted vegetables on lighter soils faces different water application requirements than a neighbor who grows deeply rooted crops in heavier soils. Furthermore, a rain which delivers two inches of water to a particular site, may deliver only a small fraction of an inch to another farm site only a few miles away, a factor which will alter demand for irrigation water.

Farmers in irrigation systems around the world are faced with the common task of hitting a moving target--a varying crop root zone moisture deficit--within an irrigation system which typically has been designed by remote engineers, managers, and politicians whose professional responsibilities were to do little more than aim a quantity of water in the general direction of a command area. In most large-scale systems, especially in Asia, the upstream control systems have been designed without adequate regard to the problems faced by farmers in securing local control (Bottrall, 1981b; Bottrall, 1985; Bromley, 1982; Kathpalia, 1981; Lowdermilk, 1986; Wade, 1979; Wade, 1980; Wade, 1982a; Wade, 1982b; Wade, 1987).

The fundamental problem is that main system managers in any system, cannot control strategic variables determining water demand and water productivity farm by farm and field by field--site specific variations in soil moisture holding capacities, soil moisture availabilities, planting times, crop varieties, root zone depths, daily crop moisture depletions, specific evapotranspiration rates, and margins to the permanent wilting points. Such matters are known to main system managers only as central tendencies, not as field-by-field particularities. On the other hand, individual farm operators cannot adequately control variables which must establish the pattern of main system water supply--watershed yields and distributions, storage and canal capacities, intra-and inter-state (provincial) allocations, river and canal hydraulics, regional or district strategies for conjunctive use of surface and groundwater, and the management of large main system storage, canal, and drainage structures. Main system supply and the multitude of farmer demands, therefore, must be matched by creating a middle level tier of organizations which accept main system *wholesale* water deliveries within the constraints which the main system must impose, control such water, and *retail* it to individual farmer users, each of whom has a somewhat unique water demand driven by local circumstances.

Reconciling Main System and Farmer Knowledge

The knowledge of irrigation officials educated in the professions who inhabit central irrigation bureaucracies, depends heavily upon generalized principles abstracted from the rich flow of natural and social processes (nomothetic knowledge). Highly-processed abstract organizing principles have pride of place in science and in the training of irrigation engineers and managers in possession of formalized knowledge of other disciplines--they are the essential prerequisite of parsimonious explanation. This general, cross-culturally viable, scientific knowledge, available to educated professionals, proceeds on the basis of rendering propositional knowledge out of particular facets of the whole system, but does not comprehend the richness of the whole; it is limited to shedding light on particular, abstracted slices of reality as in economic supply and demand curves, cost-benefit ratios, pounds of pressure per square inch, yield responses to fertilizer, channel hydraulics, sedimentation and scouring, capillary action, soil intake dynamics, evapotranspiration processes, and administrative notions of span and control. Sciences in general, and the irrigation sciences in particular, abstract general rules to construct logically connected sets of propositions about relationships among phenomena. These abstracted propositions are employed in central planning units in the design and operation of those parts of the irrigation system under the management of the central bureaucracy.

On the other hand, there is extensive local idiographic knowledge, built up over long experience, and encoded in tradition and custom in the possession of local people who may have had little access to the world of scientifically processed knowledge. Theirs is the knowledge of unique local site specific circumstances and of their particular situation relative to those circumstances. Whereas the central bureaucratic analyst must grasp general tendencies across broad systems, the farmer is intensely interested in specific outcomes on his or her particular farm. Whereas the central manager obtains knowledge for decision-making by employing methodological devices to control extraneous variables which might confuse the analysis of central tendencies in the system, the farmer responds to the very factors excluded by central management, but which importantly operate in specific local contexts. Irrigation is practiced under a great variety of conditions (e.g., social, economic, topographic, soils, climatic, and crop). These vary within a farm, they vary widely among farms, and among command areas within an irrigation system. Given the distinctiveness of each setting, the fact that each represents a unique arrangement of the generalizable properties known by central management, what seems to be a condition across the whole system is not necessarily--or even likely--a condition in any specific sub-set of that system. Farmers, employers of rich idiographic knowledge, have much reason to distrust the nomothetic understandings of main system managers.

The problem, then, is that the generalizations of irrigation managers in large, remote bureaucracies are not legitimate where farmers' individual and unique settings are concerned. The lack of mutual understanding is rooted in differences in types of knowledge and experience. There need be no hypothesis of irrationality or ill will on the part of any party to account for fundamental differences in orientation.

Public Goods and Individual Rationality

Main system managers control water by providing a transport system for water using rivers, canals, reservoirs, and diversion structures. They have assumed that if water was moved in the direction of targeted cultivatable command areas, water control at the local level would automatically evolve, simply because it was needed. In the light of history, this optimism is now known to have been naive. In the light of pressing needs for increased production and social equity, this optimism has been badly misplaced.

Local organizations to provide an interface between main and farm systems do not evolve simply because they are needed. It is necessary to understand that individually rational people, who fully comprehend the need to organize collectively to provide themselves water supply and control will, under typical conditions, not do so. This is because individual and collective rationality are different and frequently mutually opposed.

There has been much discussion of the logic of collective action during the course of the last two decades (Olson, 1965; Frolick & Oppenheimer, 1970; Mueller, 1979; Blair & Pollack, 1983; Frohock, 1987). Some have applied this reasoning directly to the problem of irrigation organization (Freeman & Lowdermilk, 1981; Lusk & Riley, 1986). The argument is straightforward. One begins by distinguishing between private and public goods (Figure 2.3). If its benefits can be captured by the investor-owner and denied to those members of the community who do not invest in it, a good is categorized as *private*. As discussed in Chapter 1, private goods are possessions such as clothing, automobiles, home appliances, personal work tools--an individual invests in them and individually enjoys benefits of ownership.

On the other hand, a good is *public* or *collective* if its benefits cannot be denied to those who do not help to bear the costs--*free riders*. Many important goods are public. Flood control projects indiscriminately benefit all subject to a rampaging river whether or not they have paid their share of the cost. A pollution control program generating cleaner air and water cannot be denied to all those in the locality who breathe and make use of the riverside but who do not pay. A tax cheater, should he or she get away with non-payment, enjoys the benefits of the common defense or street lighting as much as the taxpayer. Herein lies the problem. In collective good situations, if individuals are allowed to exercise their own rationality in independent fashion, the logic of the individually rational utility seeker is not congruent with the logic of the community. The conflict in logics can be easily illustrated.

If, for example, farmers individually observe that their leaky and mis-aligned watercourse requires improvement they will not, on individually rational grounds, invest in corrective action. Assuming a sizeable number of farmers, each will calculate as follows: if one should make the investment of time, energy, and money required to improve the fraction of the channel going through one's own land, and other farmers do not make comparable corrective investments in a coordinated fashion, then the payoff in improved water supply and control--the public good--is negligible. On the other hand, if many others undertake the improvement effort on each of their sections, and our individually rational decision-maker does not do so, he will still enjoy a substantial share of the ben-

efit provided by the work of others at no personal cost. Therefore, the rational individual will choose to do nothing either way.
The collective good will not be provided, even assuming that individuals in question possess full and accurate information about
the potential benefits, the required know-how, and resources.

This situation can be changed by establishing an effective organization which will insure that the contributions for the provision of a given public good are predictably obtained from all
beneficiaries through use of some set of enforceable joint agreements defining a *fair share* of contributions and benefits. Obligations to bear costs must be tightly interconnected with delivery
of benefit. If individuals have reason to believe that the organization will deliver its benefits without regard to member investment, then incentive to bear obligation is degraded; it becomes
rational to be a free rider, and provision of the collective good
will be compromised. Or, if the collective good is provided by
an outside altruist--i.e., a unit of government or a charitable or-
ganization--the collective good will be allowed to deteriorate as
everyone individually chooses to take a *free ride* to their short run
advantage but at the expense of allowing the public good to deteriorate for the community in the longer term. Organizations,
scaled to manage the required collective good and designed to
control *free riders* by carefully connecting delivery of the good with
fulfillment of membership obligation can defeat the individual
rational logics and can make possible local development--including
irrigation development.

Levels of Social Organization
and the Changing Meaning of Water Control

It is now possible to generate a synthesis of the irrigation problem
by employing concepts of water control, the distinction between
nomothetic and idiographic knowledge, and an appreciation of
water supply and control as a public good.

In typical large gravity flow irrigation systems, public bureaucracies
build and manage main system works. *Lumpy* public goods such
as high dams, large reservoirs, and major canals, cannot be provided by small local organizations. At the main system level, good
water management means controlling the flow of water volumes
in large-scale capital works within rather narrow parameters so
that water moves predictably toward aggregated demands of many
farmers. The emphasis is upon dealing with farmers in categories by focusing upon average needs and conditions. Main system managers everywhere work in civil service systems, and are

not sanctioned in a manner directly connected with farm productivity of the water they manage. Main system operators, while adapting their system to general features of local topography and to local histories of demand, depend heavily upon the processed disciplinary nomothetic knowledge of engineering, public administration, and economics without having the time or the need to know specific local details of individual farms. Water control at the main system level means managing water for daily operational smoothness so as to avoid sharp fluctuations which would threaten to exceed physical limits of systems operating to carry large flows within narrow tolerances.

Yet, as water flows from rivers and reservoirs toward its end-point of use through primary, secondary, tertiary, and quaternary canals, it maintains its nature as a collective good as it is divided into ever smaller flows. When it reaches the farm gate it is quickly transformed into a private good which can easily be denied to free riders. But, prior to its arrival at that point, it requires collective management.

Appropriately scaled middle level organizations, fitted with tools permitting the measurement, division, and control of water in reaches below those effectively administered by the main system, com-

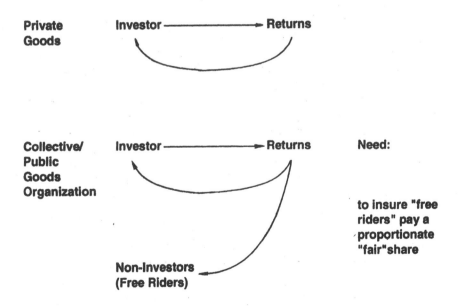

FIGURE 2.3 Returns on Private and Collective Goods.

bining general nomothetic principles with increasing amounts of local idiographic knowledge capable of effectively controlling free riders and delivering water to the farmer in a predictable and controllable manner, have not typically been made a priority by main system managements in most third world countries (Jain, Krishnamurthy, & Tripathi, 1985; Owens & Shaw, 1972; Coward, 1986a; Coward, 1986b; Coward, 1987; Esman & Uphoff, 1984; Whyte & Boynton, 1983). Such organization is found, of course, in traditional systems--especially communal systems--where traditional irrigation behavior has not been seriously disrupted (Hunt & Hunt, 1976; De Los Reyes, 1980; Korten, 1982; Martin & Yoder, 1983; Bray, 1986; Coward, 1980; Coward, 1986b). Also, such organization is found in the affluent nations of North America, Europe, and Northeast Asia (Rangeley, 1987; Bray, 1986; Maass & Anderson, 1986). The successful functioning in Japan of local farmer irrigation organizations and their transfer to Taiwan and Korea under the auspices of Japanese colonialism has been a well documented story (Kelly, 1982a; Kelly, 1982b; Bray, 1986). Furthermore, devolution of water management through use of local organizations has been demonstrated to be a key to efficiency and equity as witnessed in the contrasting cases of the Meiquan system in the Peoples Republic of China (Nickum, 1974; Nickum, 1980) and the Dhabi Kalan in Haryana, India (VanderVelde, 1980).

But, many centrally managed systems of the Third World constructed in the rush of the nineteenth and twentieth century expansion, administered by elites who have largely stood outside the traditional local organization of their own societies, have not sufficiently attended to preservation or promotion of middle level irrigation organizations (Coward, 1980; Keller, 1987). Lack of decentralized local organizations has been acknowledged to adversely affect irrigation projects in South Asia (Vaidyanathan, 1983; Chambers, 1980). Given lack of effective autonomous middle level local organization, main system managers have been forced to administer water flows further and further downstream where their concern with keeping water flows smooth to accommodate main system agendas are inappropriate in the face of local farmer desires for rapid adaptation.

Good water management at the farm level (Figure 2.1) must focus on controlling water such that relatively small volumes are productively placed in particular crop root zones in specific and unique individual settings. Water must be moved at the proper time, in the required amount, such that the micro-environment of the plant is conducive to maximum production. Controlling water to provide proper micro-environments for crops requires a great deal of skilled labor. Above all, farmers must rapidly adapt to field specific changes in demand; they confront various crops planted in different soils, at different stages of growth, under variable weather conditions, all the while bonded to expectations of dynamic kinship networks and neighbors. The compelling fact at the farm level, is that delay of a water issue for three or four days under quite typical conditions of crop, soil and climate at a critical period in plant growth can cause severe decrements in plant yields. As farmers witness their plants moving toward the permanent wilting point, they actively seek ways to obtain water, authorized or not, from the main system. Farmers are not persuaded to act on central tendencies in the irrigation system, but must be attentive to unique conditions of particular fields and crops--factors that main system management cannot attend to without standing in violation of main system agendas. Farmers cannot depend heavily upon processed disciplinary knowledge except as it is adjusted to their particular situations. Unlike main system managers, farmers are directly rewarded and punished according to the productivity of water.

In the absence of effective interfacing organizations to reconcile main system management of supply with farmer demands, lower level main system managers are generally faced with an impossible choice: (1) to maintain as much distance as possible from local patterns of privilege and free riding in pursuit of controllable water, or (2) to become entangled in countless energy absorbing local conflicts, complaints, and specific demands for which their training, knowledge, and organizational resources are grossly inadequate. If main system managers are left in the local area sufficiently long to become deeply knowledgeable of local circumstances, they tend to become attached to local power alliances defending existing distributions of advantage. Even if such problematic patterns could be altered by intervention of main system managers, it would only institute a somewhat different pattern to the advantage of other sub-optimizers who, just as quickly, would undercut the collectively provided water supply and control for the whole. The distribution of misery would shift from group to group but the total quantity of misery would remain little changed. If one transfers main system personnel regularly to prevent emergence of local attachments, adequate local knowledge and linkage must be sacrificed and small groups of local irrigators will be no less free to arrange whatever pattern of local advantage available to them, possibly at considerable cost to overall system functioning.

At some point, the level of water disaggregation becomes sufficiently close to the farm, too remote from main system knowledge of central tendencies, and too close to the rich array of local crop consumptive demands to be effectively managed by the main system. At this point, local farmers must come into play in an organized way. There are, fundamentally, two options: (1) design an organization which guides farmer participation so as to protect both main system supply agendas and farmer demand requirements, or (2) allow farmer participation to take the form of whatever alliances opportunistically emerge to join with lower echelons of the main system bureaucracy. Farmers will participate, for better or for worse. Irrigation bureaucracies vary widely in their approaches to this problem of managing the transition from state bureaucracy to farm, but in every instance some form of routine, organized interaction emerges, to manage the control of water as it is progressively divided into ever smaller volumes. Knowledge of specific conditions beyond the awareness of main system managers must be brought to focus on the precise control of water in the downstream reaches of canal systems.

Effective organization at the middle level is essential to provide linkage between farm water demands and main system supply. This requires jointly negotiated agreements with main system operators and among neighboring irrigators. These agreements, written or unwritten, formal or informal, are the *stuff* of organization. The question is not whether such agreements appear at the middle level, but whether a given set of joint agreements serve any defensible conception of irrigated agricultural development and social equity. If conscious organizational effort is not undertaken with both farmer and main system support, emergent middle level organizational agreements will reflect individual sub-optimum free riding rationalities, not arrangements which best serve the potential community of irrigators. The next two sections will present a discussion of what joint agreements need to be made to create middle level organizations.

Components of Organizational Design for Water Control: The Interface Between Main and Farm Systems

If one accepts that there are different requirements for water control and knowledge as between main and farm systems, and that effective middle level organizations must link and reconcile otherwise incompatible irrigation agendas to provide water supply and control as a public good to farmers, then it becomes necessary to design organizations to harness and discipline farmer participation to the operation of middle level organizations functioning in the interface between main and farm systems. Local irrigation organizations are assemblies of joint agreements among farmers and main system managers which make possible the production of collective goods not available through individual effort. If the organization can produce and distribute such a collective good--sufficient water supply and control--to its members while denying it to *free riders*, then members will pay the organizational costs of supplying and controlling water, i.e., the costs of allocation, maintenance, and conflict management. Joint agreements that compose a middle level organization include: (1) joint agreements about the direction of staff authority, (2) joint agreements about patterns of staff recruitment, and (3) joint agreements about mobilizing resources, distributing water via share systems, and connecting maintenance to allocation. The essential choices defining the nature of optional organizational joint agreements are outlined in Figure 2.4 which states the essential working hypotheses about which agreement options are thought to earn farmer endorsement of the organization.

Staff Authority Relationships

A critical variable in middle level organizations is that of establishing authority relationships. Shall the staff of the local organization be fundamentally responsible to authorities upstream in the main system or shall they be directly accountable to farmers? Responsibility to main system authority is indicated by dependence upon the main system for renumeration and by affiliation with managers. Responsibility to farmers, a posture of looking downward in the system, is typically indicated by the fact that farmers can hire and dismiss organizational staff without regard to civil service regulations, and by the fact that rewards for services are established by farmers. It is hypothesized that as staff of the middle level organization *look down* for direction and definition of success, they become tuned to the local requirements and acquire incentive to creatively seek methods to fulfill local water demand within main system constraints. As staff *look up* to main system authority for definition of job performance adequacy, they become local agents of main system principles without substantial incentive to be concerned in sustained ways with seeking the most workable fusions of nomothetic and idiographic understandings in the service of water control in local reaches of systems.

Staff Recruitment

A middle level irrigation organization may be staffed by *cosmopolitans* who are recruited from outside the local command area, who are typically selected on the basis of educational qualifications with a high emphasis on comprehension of nomothetic generalizations of a given discipline frequently operationalized in the form of examinations, who are characterized by considerable social distance from local farmers, and whose career aspirations are for upward mobility and departure from the local irrigation command. On the other hand, the staffing choices may emphasize the hiring of *locals* who are recruited on the local labor market by employing criteria having to do with local experience, local social connections, virtually no social distance from the farmers being served, and aspirations to spend a lifetime of work in the local command area. The greater the proportion of staff recruited on the *local* pattern, the greater the propensity for staff to integrate idiographic understandings of the particular setting into implementation of main system operational requirements established upstream to serve local water control needs.

Water Share Systems

· Water share systems are central to the life of viable irrigation organizations, and appropriately designed joint agreements about the collective use of rules and tools to establish membership, capture water from the upstream main system, allocate water to farmer demand, and mobilize esources for paying the organizational costs of water management, stand at the very heart of effective local organization.

Effective joint agreements about means to allocate water and system maintenance center on the concept of *water share*. A water share is always a two-sided concept: (1) It confers legitimate access to the water resource within certain pre-arranged rules; and (2) it imposes upon the user a specified obligation to share in payment of the water management costs. The concept of share unites two essential aspects of organizational operations--resource allocation and acquisition. Productive and equitable water distribution is not a matter of good intentions; it is primarily a function of the way organizational rules resolve the problem of defining and allocating water shares. Even though models are now becoming available to probe aspects of the problem (Molden, 1987), the subject is complex and no comprehensive analysis of the problem has been performed. It is only possible to make brief mention of strategic considerations and issue a call for sustained cross-cul-

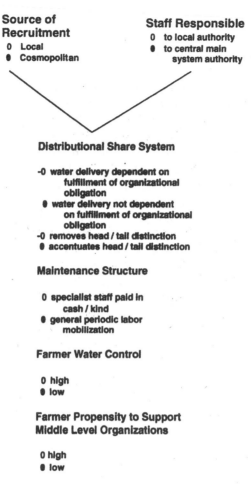

Source of Recruitment
0 Local
● Cosmopolitan

Staff Responsible
0 to local authority
● to central main
   system authority

Distributional Share System

-0 water delivery dependent on
   fulfillment of organizational
   obligation
● water delivery not dependent
   on fulfillment of organizational
   obligation
-0 removes head / tail distinction
● accentuates head / tail distinction

Maintenance Structure

0 specialist staff paid in
   cash / kind
● general periodic labor
   mobilization

Farmer Water Control

0 high
● low

**Farmer Propensity to Support
Middle Level Organizations**

0 high
● low

FIGURE 2.4 Staff Recruitment and Responsibility.

tural investigation of the problem. This discussion of water distribution systems has been influenced by that proposed by Maass and Anderson (1978; 1986). Essentially, middle level organizations can specify water shares in some combination of the following ways:

Shares may be organized by fixed percentage allotments: (1) by volume (e.g., a percentage of the total acre feet or cubic meters estimated to be available); and (2) by time period rotation (e.g., a percentage of a day or week).

Shares may be organized by a priority system: (1) priority by location (e.g., head to tail of a channel); (2) priority by farm characteristic (e.g., time of settlement); and (3) priority by crop (e.g., market or subsistence value).

Shares may be organized by user demand: (1) demand placed upon storage in a surface reservoir; and (2) demand placed upon storage of groundwater.

Many combinations of distributional principles are possible, depending on local circumstance. Share systems may be combined by constraining one type of share by another. For example, shares by volume may be subjected to crop priorities. Distributional systems may employ two share types simultaneously as when shares by time rotation are supplemented by higher priced demand water. Share systems may shift within seasons in response to change in the environment as in cases where shares by volume are shifted to shares by time of settlement or crop priority during severe drought. The great diversity of irrigation allocation arrangements observed around the world represent various combinations of these basic distributional principles, e.g., the *warabandi* systems of the Pakistan and Indian Punjab (Reidinger, 1974; Kathpalia, 1981) are combinations of distribution by percentage of weekly time period combined with priority by location. This is to say that water is run from the channel head to tail and allocated to farmers for a time period calculated to be proportionate to area as follows:

Hours per farm	Area per farm
168 hours/week	total area in water
course command	

Whatever water runs in the channel during that time period goes through the channel outlet to the farm.

A case study has recently been prepared which reports, in detail, the de jure and de facto operation of the *warabandi* (rotational) share system at the Niazbeg site in Pakistan (Shinn & Freeman, 1988). Given lack of an appropriately designed middle level organization to administer the system, water control is low and crop potentials are far from being fulfilled. An Indian case study of a small tank system in Madhya Pradesh reports that the de jure design calls for a demand contract system on reservoir water but, given a lack of physical means to control water delivery and lack of effective organization to connect water delivery with local water assessments, farmers and main system management have experienced much difficulty (Bhandarkar & Freeman, 1988). At Lam Chamuak in Northeast Thailand, the system has fallen into disarray due to an absence of a viable share system which could unite main system management and farmers around agreed-upon and enforceable joint agreements regarding water allocation and maintenance responsibilities (Parnakian, Laitos, & Freeman, 1988). A water user association exists in a pro forma sense at Lam Chamuak, but its design did not incorporate a viable share system. In two tank systems located in the Sri Lankan dry zone (Wilkens-Wells, Wilkens-Wells, & Freeman, 1988), the case study reports that no effective share system has been devised for management of main system controlled distributaries in the lower reaches of the commands. However the *vel vedane* works with varying success in administering a rotational share system on channels ostensibly controlled by the main system but, in fact, controlled by farmers. Variation in the effectiveness of administration of the distributional share system was found to substantially impact water control, yields, and farmer willingness to support the local organizational arrangements.

Martin and Yoder (1983) have presented a case study of two Nepali systems--Chherlung and Argali--which makes clear how farmers have successfully created two distinctly different organizations around purchased water share arrangements, each of which closely tie water delivery to farmer payment of assessments. In Argali, farmers devised a system of sharing water by allocating channel flow in proportion to the area irrigated, whereas, at Chherlung, water was shared out by fixed proportions of flow volumes which could be detached from any given piece of land. The latter was observed to provide an incentive to increase water use efficiency, as incentives existed to obtain the greatest possible productive value of each unit of water so that water saved could be transferred to other irrigators with deficient supplies and who were willing to pay for it.

Water Distribution Systems and Water Control

Problems of water control will assume different forms, depending upon which combination of share distributional principles are employed to manage the resource from the main system to the middle level and, in turn, from the middle level organization to the farm gate. Different kinds of middle level organizational problems emerge to confront farmers and managers, depending upon choices made in the establishment of water shares.

Three facets of the problem can be abstracted out of the complexity (Freeman, 1988): (1) the matter of defining membership in the irrigation organization, (2) the issue of connecting water supply and control with member fulfillment of obligation to the organization, and (3) the matter of addressing the impact of head and tail location in the system.

First, to be a member of the irrigation community, one must be defined as such by some legitimate organizational principle associated with definition of water shares. One does not become a member of an irrigation community simply by living in an area proximate to canal flows. Each water distribution system, emergent out of some combination of the principles outlined above, specifies a means to be defined as a member. For example, in a typical Indian or Pakistani *warabandi* system--one built on the principle of allocating a share of the 168 hour week in direct proportion to the proportion of cultivatable land owned and/or operated by the irrigator in the approved command area--one becomes a member of the irrigation community by virtue of owning or operating cultivatable land within the approved command area. However, if one operates a local irrigation organization on the principle that proportions of investment (purchase of shares) in the organization can be made without regard to acreage, then ownership of organizational shares or *stock* define organizational membership, e.g., the case at Chherlung in Nepal (Martin & Yoder, 1983), in the case of the Philippines (Coward, 1985) and in some systems in Spain and Colorado (Maass & Anderson, 1986). Therefore, joint agreements about distributional shares become agreements about who is, or is not, a member of the irrigation organization.

Secondly, creators of water distributional systems must confront another strategic choice--how closely connected will water service be to fulfillment of organizational obligations. Farmers anywhere in the world cannot pay to *own* water; virtually everywhere water ownership is retained by the state for public purpose. Yet, everywhere water supply and control exacts costs lower in the system just as it does in the upstream reaches under the jurisdiction of main system management however defined. Controlling water on tertiary or quaternary channels closer to specific farmer demand schedules may present problems substantially different from managing water in a given set of primary and secondary canals, but costs of water control must still be paid--costs of personnel, measurement and division devices, channel maintenance, conflict management. If one is going to design a program for recruiting local people who will be responsible to the farmer irrigation community, then the local organization must pay, if not all, at least a significant fraction of their wages and salaries or lose control over the ability to hire and dismiss such personnel and to define the very nature of their job priorities. Therefore, middle level organizations do not raise resources to *own* water, but resources must be mobilized to pay operation and maintenance costs for managing water under the organization's jurisdiction. Two strategic questions arise:

(1) Are organizational joint agreements about water share systems established such that water service is directly dependent upon member payment of that member's share of the cost? Or, is water delivery divorced from fulfillment of member's cost obligations?

(2) Are farmer shares of management costs at least roughly proportionate to water service received?

Regarding the first question, patterns of water management observed at four case study sites (Freeman, 1988) establish that nowhere is there a close connection between water delivery and farmer payment. In the Pakistan case, farmers pay an assessment based upon crop type and estimated yield, and they make such payments in a manner totally unrelated to water supply or control received at the farm gate. Therefore, a farmer receiving relatively good canal water service pays according to the same assessment schedule as one who receives relatively poor service, unless one simply refuses to pay. A water share system designed around delivering, in a fixed way, whatever highly variable volume of water happens to be flowing during the farmer's weekly time allotment and which largely divorces charges from the water service does not earn the enthusiasm of the more disadvantaged irrigators. In the Indian, Thai, and Sri Lankan cases, water service and fee collection are also divorced. In each of the case study systems, those who fail to pay their assessments are not meaningfully penalized. Free riding is the norm. In the Indian case, uncollected revenues have mounted to considerable sums. Farmers are quick to see that, from an individually rational standpoint, one is rather foolish to pay assessments--especially those whose water supply and control is decidedly inferior. To disconnect farmer payment of assessments, whether in cash or kind, from water delivery is to virtually invite free ridership and organizational decay.

With regard to the closely related second question, farmers are intensely interested in having their water assessments reflect something of the amount or proportion of water obtained. A share system which connects a conception of variation in assessment to variation in supply is likely to earn greater enthusiasm from farmers than one which does not do so. For example, in a Pakistan case study of the Niazbeg distributary (Shinn & Freeman, 1988), there is a strong inverse relationship between water cost and control. Given a canal system operated under fixed rotational system, and given that irrigation revenue assessments have been disconnected from actual water deliveries, the more disadvantaged farmers located toward the tail pay much more per unit of water delivered than those advantaged by head locations where water supplies are greater. The cost of tubewell water in the command area was found to be roughly 340 percent more than the cost of canal water, but not only was it worth more due to its high controllability, but there was a direct relationship between amount of payment and amount of water received.

A third aspect of the distributional problem can be stated as a question: does the share system reinforce or resolve the problem of *head* and *tail* location given by geography? Water must flow in channels from point A to B. Farmers toward point B are, by definition, nearer to the tail and--all else equal--will be disadvantaged in the matter of receiving water allocations relative to those increasingly near point A. The more one proceeds to locations toward the tail of an irrigation channel, the more one is vulnerable to: (1) losses due to leaks, seepage, and evaporation; (2) self interested manipulations of others toward the head as the number of irrigators intervening between farmer X and the head increases; and (3) non-routine breakdowns in the system--there is simply more to go wrong when one is dependent upon longer channels.

Engineers must construct canals with potentials for head and tail positions, but it is up to designers of social organizations to decide whether the potentials for head-tail distinctions are to be realized by organizational share systems.

The Pakistan, Indian, Sri Lankan, and Thai rotational water delivery systems accept, reinforce, and solidify the head-tail distinction. When one combines water allocations by time and location, one immediately reinforces what engineers and geography have already done, and one creates a fundamental difference in interest between irrigators at the head and tail positions which must threaten the solidarity of any local farmer organization. Irrigators toward the head do not experience the same water supply and control problems faced by their neighbors located toward the tail, and they typically find their relatively advantageous situations to be threatened by the desires for reform on the part of tail-enders. Tail demands for more water, and more timely water, appear to come at the expense of head-enders advantage who tend to show less interest in solving problems for tail-enders.

Water supplies at tail positions are, however, a problem only to the extent that organizational design of the share system fails to overcome them. If the middle level organization should employ a combination of distributional principles which impose the costs of *water loss* or *shrink* on all members without respect to location, then all members have equal incentive to pay costs of maintenance and operation of the system as a whole. If channel losses anywhere are distributed by the share system to all, irrigators at all locations have equal concern to reduce losses at any point. If, for example, the organization should distribute water by volume, or by volume combined with some form of demand, and if volumes are measured so that losses anywhere on the common channel reduce volumes to all irrigators, and if assessments against shares are proportionate to volume received, then all farmers absorb the *shrink* and all have an incentive to reduce losses. Views of *head* and *tail* as an inevitable natural phenomenon must be set aside. Uncritical willingness to accept inequities in irrigation commands is a function of poor organizational analysis of share types, and is not a reflection of universal physical necessity.

One cannot, within the confines of this chapter, perform an analysis of the complexities of selecting and defining the physical tools and organizational rules for implementing distributional shares. However, three critical questions must be raised when designing organizations: (1) What are the implications of a given distributional system for defining membership in the irrigation community, (2) is water delivery connected or disconnected from irrigator fulfillment of assessment obligations so as to control free riders, and (3) does the particular distributional system actualize the negative potentials for problematic water control inherent in a reinforcement of the head-tail distinction among farmers, or does the share type distribute the water losses to all members without respect to location?

Maintenance

An important subset of the general distributional problem has to do with how routine maintenance is organized. There are at least two strategic options.

The first option is that routine maintenance can be performed by staff hired full- or part-time, paid in cash or kind by resources mobilized by member water share obligations, and thereby develop specialized competencies in aspects of irrigation maintenance work. Or, as a second alternative, tasks may be performed periodically by mobilizing farmers, or their surrogates, who perform maintenance work within specified time periods or be subject to penalty.

It is hypothesized that as routine maintenance is performed by specialized and paid staff as employees of the local organization, water control closer to the farm gate will be enhanced because:

(1) Individuals who labor full or part-time in such work acquire specialized skills, job related contacts in the irrigation community and marketplaces, and special knowledge not available to the general farm membership not developed by annual or semi-annual general labor mobilizations.

(2) Full-time paid staff can promptly respond to problems in the command area whereas periodic labor mobilization schemes tend to defer to slack seasons much routine maintenance too large in scope for the voluntary efforts of any few farmers. Obviously, farmers will quickly set aside their personal farming agendas to mobilize when their system is seriously threatened by emergencies, but the general pattern of intermittent labor mobilization does not place a priority on constant, careful, detailed attention to common maintenance problems emerging incrementally.

(3) The system of general labor mobilization provides much opportunity, even incentive, for organizational *free riding*. Farmers who are *free riders* may find it in their interest to schedule other activities during the time that labor is to be mobilized for maintenance so as to secure benefits of maintenance without contributing a *fair* share of work, however defined. The organization is then on the defensive; it must proceed against *free riders* in ways which threaten to erode support for the organization or at the very least impose substantial costs on the organization. Costs will be high because those with sufficient influence and power to attempt *free riding* strategies are those who are most difficult to keep harnessed to organizational norms. It is thought to be much less disruptive to the organization to collect operational and maintenance revenue according to some legitimate conception of water shares and to use the payments in cash or kind to support a continuously employed full- or part-time maintenance staff.

Farmer Propensity to Support Local Organizations

The final variable on Figure 2.4 is that of farmer propensity to support local, middle level organizational arrangements interfacing farm and main systems. Support here is taken to mean a willingness to: (1) make investment of personal resources to sustain the distributional arrangements for controlling water, and (2) abide by organizational rules. It is posited that farmers are willing to make such investments, and to accept organizationally imposed regulations, insofar as their water control requirements are at least minimally fulfilled. For this to occur, the middle level organization must provide an arena of security and predictability within which farmers can count upon: (1) organizational agreements about allocation, maintenance, and conflict resolution being enforced; (2) their assessment revenues being locally spent on local water supply and control problems which they experience; and (3) water delivered so as to fulfill their crop consumptive demands. Evidence was collected on farmer propensity to support local organizations in Pakistan (Shinn & Freeman, 1988), Thailand (Parnakian et al., 1988), and Sri Lanka (Wilkens-Wells et al., 1988), and farmers in each case evidenced strong desire for improved organizational arrangements and a willingness to give their support by way of payment and loyalty, if such organizations providing effective water supply and control could be developed.

The argument can now be summarized: the more the middle level organization is staffed by *locals* who look downward in the system to the authority of farmers, and the more it provides continuous maintenance performed by employees using a system of water shares which denies water to *free riders* and distributes the water shrink to all members without regard to location, the greater the water supply and control that can be afforded across the system, and the better the opportunity for farmer involvement and investment. Farmers will display a higher propensity to deliver sustained support to such an organization which can adjust the general requirements and central tendencies of the main system water control agendas to locally comprehended water control needs.

Assembling the Components of Middle Level Organizations

Components of effective interfacing organizations have now been specified. Of course, no mere discussion of items represents a real organization any more than a listing of building materials represents a house. The properties of an effective functioning organization are considerably more than a list of the parts. Furthermore, there is no universal blueprint for assembly of the component parts in the domain of local organizations any more than there can be universal blueprints for assembly of bolts and braces into bridges. There is no cause for despair in this observation for at least two reasons. First, farmers in irrigation systems are capable of building new or improved organizations adapted to site specific circumstances if they are given support to do so. The problem is less one of farmer capability and more one of recognition of need and provision of the required balance of constraint and autonomy, incentive and direction. Second, there are general principles which can be advanced to guide organization building, even if simplistic recipes are of little use.

Researchers involved in each of the case studies shared one common experience. Key informants and sample farmers, at each of the studied sites, were not awed by the thought of building new or improved farmer managed organizations to locally supply and control irrigation water where their skills and resources were seen to be adequate for the collective tasks. At each of the case study sites, at least some farmers had been discussing the problem for years.

Farmers, in general, do not have to be convinced of the wisdom of getting organized to improve their water supplies and control, of creating a locally organized zone of security for their local understandings and resources to be applied to their water control problems, nor do they generally resist thoughts of hiring local people with their own resources who would be subjected to farmer water management priorities. Nor do they resist the idea of getting organizational leverage over violators of rules and abusers of tools by making water delivery conditional upon fulfillment of organizational obligations. Farmers indicated an understanding that water service must somehow be connected to gathering in resources for running water. Farmers seek predictability and control over a vital resource and they are, given what they believe to be a credible arrangement, willing to organize to get water control and keep it. Yet, farmers are skeptical that local influentials may subvert local organizations to their private purposes, and that authorities in the wider environment may not give sustained support to organizations if such support should conflict with well connected influentials. They tend to be concerned that, whereas local elites may be allowed latitude by main system managements to whipsaw local organizational personnel and other resources, that non-elite farmers might not be allowed latitude necessary to organize themselves in ways seen practical for themselves. Irrigators generally see problems with investing their resources in systems where they possess no officially sanctioned roles or responsibilities. Why tax oneself for a collective effort on systems where it is not even supposed to officially function and which can easily fall prey to the interests of local elites and main system managers?

Therefore, it is necessary to briefly examine conditions under which chances improve for assembling the necessary components of local farmer organization in a manner acceptable to farmers and main system managers. The discussion will revolve around three topics: (1) basic design premises, (2) fundamental conditions which must be fulfilled prior to any farmer organizing effort, and (3) an outline of general organizational structure and processes.

Fundamental Design Premises

(1) In the implementation of any organization building process, experience gained by the successes and failures of farmer participatory approaches tested in several nations must be employed and integrated with the sense of organizational design advanced here. Much has been learned about promoting farmer participation in development projects generally and in irrigation systems particularly (cf. Bagadion & Korten, 1985; Cernea, 1983; Cernea, 1985; de Silva, 1981; FAO, 1985; Illo & Chiong-Javier, 1983; Korten, 1982; Freeman & Lowdermilk, 1985; Lynch, 1985; Montgomery, 1983; Moris, 1981; Siy, 1982; Uphoff, 1986a). This literature will not be reviewed here, but it will suffice to note that a sense of organizational design remains inert unless farmers are activated to give shape and life to the design components. Attempts to promote farmer participation without a clear sense of organizational design push unorganized farmer demands up to main system management. Given its different supply agenda, the main system cannot cope with the variety of farmer demands. Meaningful farmer participation requires that an organizational framework be designed to focus participation on performance of specified responsibilities in particular reaches of the system. Design without participation is an empty exercise; participation without careful organizational design is futile for farmers and threatening to main system managers.

(2) The issue of designing organizations must be viewed in the legal context appropriate to the specific site. Obviously, much can be said about the manner in which different legal traditions cast the problem of water rights and responsiblities (Radosevich, 1986; Radosevich, 1987). Given the complexities of legal reasoning about water, one must advance generalizations with caution. One generalization can be advanced, however, which enjoys broad validity--nation-states generally reserve ownership of water for the public domain. Such public legal ownership of the water does not pose an obstacle to farmer organization because ownership of water per se is not a requirement of effective local organization. The issue centers on defining responsibilities for management of publicly owned water and organization of means to pay the costs of management. Farmers may not be able to *own* water in a legal sense, nor buy or sell water per se, but they must organize to manage water which the public domain has placed in their

trust and pay the local costs of management. When farmers around the world organize, they purchase water control rendered out of good management; they do not purchase water per se. It is possible for farmers to *own* some segment of the facilities for water control, and to collectively *buy* water control without *owning* water.

(3) The issue in linking local water user organizations to main system managements is less an issue of centralization versus decentralization and more an issue regarding how to *devolve* responsibilities to the local command area organizational unit within parameters acceptable to higher authority and subject to specified oversight of main system management (Esman & Uphoff, 1984; Montgomery, 1974). There is little promise in organizational visions that assume the local farmers' organization is simply an extension of main system bureaucratic management, nor in a vision of decentralized farmers' organizations working to obtain water from the main system in whatever opportunistic ways are made available by local circumstance. Devolution of water responsibility means that within standards and criteria established by the main system and periodically reviewed by higher authority, the local organization is empowered to act as an autonomous or quasi-autonomous unit with its own control to operate with its personnel, budget, and management procedures. Should the local unit fall into significant violation of its mandate--either by allowing physical tools to deteriorate or by being taken over by local forces, unsupportive of its mission to equitably serve all the farmers in the command area as stipulated by the charter--then the main system management must exercise its one meaningful sanction, the threat to withdraw water supply in proportions appropriate to the nature of the problem.

Conditions Precedent

Prior to initiating organizational design with authentic farmer involvement, political authorities and administrative managers of the main irrigation system must have arranged to recognize and support farmer water user organizations. Bagadion and Korten (1985) summarize the need for main system administrative support in the context of promoting farmer involvement. At the very least, the following conditions must be fulfilled as a condition of proceeding with organizational design:

(1) Political and administrative authorities responsible for main system irrigation must legally recognize the existence of local farmer water management organizations. There must be clear agreement that, at a given point in the system, specified farmer organizations will accept responsibility for management and control of water

allocation, maintenance, and specified conflict management activities. Without such administratively recognized and legally enforceable recognition of the local organization, *free riders* will be able to exploit unresolved definitions of responsibility and thereby make it difficult for either local leaderships or main system managers to exert control of behavior threatening farmers who do meet their organizational obligations. If free riding becomes a successful strategy, incentive is quickly lost for all farmers to contribute to the provision of the collective good.

(2) Farmers, in addition to paying any property, income, or other taxes imposed by the state, must be permitted to raise their own revenue through their organization's distributional share system. This revenue must be locally managed and invested to mitigate water supply and control problems as defined by the local organization. Annual or seasonal costs of running the local reach of the system must be totaled and assessed to water user members in a manner somehow proportionate to water delivery service received as defined by the organization's share system. Revenues are to be raised in cash or kind to cover costs of hiring local staff, purchasing materials, and hiring local contractors and temporary labor. It is essential that water delivery to each farmer water user be directly dependent upon payment of organizational assessments.

(3) Main system authorities must be prepared to support operations of local command area organizations in local command areas as they exert control upon free riders. Main system authorities, after having endorsed a charter and by-laws specifying operational procedures for local water user associations, must be willing to uphold judgments made in accordance with those procedures. If free riders learn that the main system management is less than firm in its support of local organizational procedures, local organizational leaderships will be placed under severe burden in attempting to control free riders who find powerful allies upstream in main system management.

Organizational Structure and Process

The essential structural form of local water user associations is diagramed in Figure 2.5. It is a form adaptable to the rich variety of diverse cultural meaning systems and the logic is straightforward. Member shareholders elect representatives to a governing board or council. The council is empowered by joint agreements with the main system and local irrigation community to direct the affairs of the local organization in accordance with the established joint agreements recorded in a charter and by-laws to which the shareholders have publicly and legally pledged themselves. Some number of board or council representatives, chosen to represent different categories of water users, meet periodically to establish policy and hire the daily operating staff to implement policy. Generally, the board or council members will be elected for two or three year (or seasonal) terms in a staggered manner.

Depending upon organizational design choices discussed earlier, the board will oversee conduct of water allocation, maintenance, and conflict management with either a local manager and staff, a main system manager and staff, or some combination of the two. Farmers will prefer locally hired staff who are fully responsible to the local organization, especially if they are paying all or a considerable fraction of the costs of staff employees. In any case, farmers serving on the board must have discretion in hiring and firing the local manager and staff without having to take their personnel cases to the main system for review and final decision. The manager and any staff, in turn, work on a full or part-time basis to:

(1) Allocate water according to the organized share distribution system;

(2) Maintain the local irrigation facilities for which the organization is responsible with resources collected from the shareholding membership according to the rules specified by the share system, and;

(3) Manage conflicts among irrigators by administering policies of the board which, if appealed, will be addressed by the board. If conflict cannot be satisfactorily resolved at that level, it would have to go to the formal legal system.

The daily needs of the local water users as defined by the particular water share distributional system are thereby served. Water users, then, in turn, elect representatives to the board, thereby completing the organizational cycle of authority.

Local organizations, composed as they are of negotiated agreements with which to conduct collective action, are the outcomes of continuous bargaining and maneuvering for advantage. Furthermore, compliance with organizational expectations must always be secured within a nested set of other relationships rooted in family and kinship organizations, credit and other supply organizations, marketing organization, religious, and political networks. Organizational structures, therefore, must possess some attributes which will permit their continuous adaptation of the organization to shifting situations. The structure of organizational rules, joint agreements for guiding the use of physical tools and structures, must possess specific attributes if the rules are to provide effective framework for organizational behavior.

65

Organizational rules having to do with staffing, exercise of authority and responsibility, and the system of water shares, must have certain properties as presented in Figure 2.6.

Analysts, farmers, and main system managers must continuously evaluate the following questions and design specific joint agreements in response to each: What kind of free riding behavior is emerging? What local organizational responses to the free riding are possible? What local organizational responses to the free riding are most effective, given local resources and water control agendas? What main system responses are possible to the free riding? What main system responses will be most efficacious, given main system water control resources and agendas? What threats to water supply and control, and, therefore, water productivity

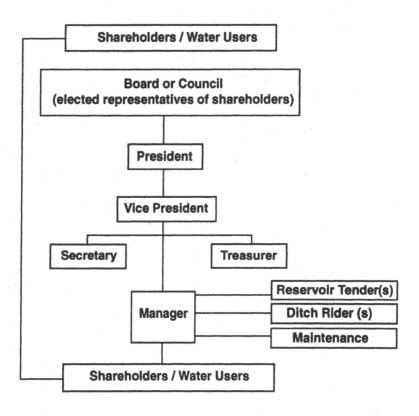

FIGURE 2.5 Structure of Water User Associations.

for farmer members, are being encountered in the main system and in the local command area? What adjustments are required in main system operation to respond? What adjustments are required in the command area internal to the local water users organization?

Continuous attention to questions such as these, combined with rapidly adaptive organizational responses with support of main system management, is the essence of successful management.

Conclusion

The meaning of irrigation water shifts dramatically as it flows from the main system, through middle level organizations to the farm. In the social division of labor within large-scale gravity flow irrigation systems, those who work with idiographic knowledge on farms are faced with fundamentally different water control problems than those in central bureaucracies who work to control great volumes of water by relying heavily on nomothetic principles and respond to demands known primarily in the form of central tendencies. One cannot assume a commonality of interest between main and farm system operators. One must assume just the opposite: namely, that farmer requirements for flexible water which hits crop moisture deficits must directly confound main system management's interest in keeping water flows smooth within narrowly defined parameters. To note that interests differ is not to suggest that mutual cooperation is not possible or desirable, but irrigation development programs designed to promote effective farmer participation in irrigation system operations must be rooted in careful analysis of the organizational link between main system and farm operations. Comparative analysis of irrigation systems may be usefully informed by examination of the variables and relationships advanced for the analysis of middle level organizations.

Cultures may vary, but fundamental cross-culturally viable forms of organizational life can be designed which can integrate local knowledge with main system knowledge, provide security for local investment against potential depredations of free riders, and integrate main system water supplies with farmer water demand. If the organization, via appropriately designed staffing patterns, directions of authority, water share systems, and maintenance capabilities, can provide its members effective water supply and control, it can secure farmer support. Farmers will support organizations that deliver payoffs in the form of water control, if the payoffs are effectively denied to free riders.

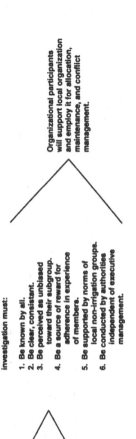

Premise: The organization delivers a service—water control at the farm gate—which pays off for the farmer-member. Water service is tightly connected to fulfillment of member obligation.

Rules must:
1. Be known by all.
2. Be clear, consistent.
3. Be perceived as unbiased toward their subgroup.
4. Be a source of reward for adherence in experience of members.
5. Be supported by norms of local groups.

Violators of rules must be quickly be identified.

Procedures of investigation must:
1. Be known by all.
2. Be clear, consistent.
3. Be perceived as unbiased toward their subgroup.
4. Be a source of reward for adherence in experience of members.
5. Be supported by norms of local non-irrigation groups.
6. Be conducted by authorities independent of executive management.

Organizational participants will support local organization and employ it for allocation, maintenance, and conflict management.

FIGURE 2.6 Properties of Organizational Rules.

There is a tendency, when the design of middle level organizational arrangements is defective, for central irrigation authorities to blame farmers for problems. Irrigation bureaucracies gravitate toward strong paternalistic behavior justified by finding farmers to be uncooperative and in need of stronger regulation to mold them in directions required of technical solutions devoid of organizational thought. As noted in Chapter 1, the consequence has been enormous centralization of bureaucratic power to manage irrigation development and to by-pass local farmer experience. As authority centralizes, communication gaps widen, local action is stifled, dependency of farmers increases, opportunities for coercion and corruption expand (Wade, 1979; Wade, 1980; Wade, 1982a), and more authority is swept up by the bureaucracy. Frustrated attempts at corrective bureaucratic action, aided and abetted by international donor agencies--speaking in terms of computerized and intensified management, coordination, monitoring, and evaluation--simply intensify the destructive spiral.

Rehabilitation of poorly performing irrigation systems is not simply a matter of reconstructing physical facilities to meet original design standards. It is a matter of examining defects in the middle level organizations to determine why they could not deliver water in sufficiently controllable ways so as to earn the support and investment of resources on the part of the irrigators. After such a determination has been made, re-designed middle level organizations may be constructed. Obtaining support from farmers for such work is not an insurmountable task if construction is combined with organizational development that realistically delivers effective water supply and control.

If there is any prospect for authentic water revolution, it will come as a result of increased water control for farmers by building improved middle level organizations whose arrangements satisfactorily link farms to state bureaucracies. What is required is a policy- relevant social science that can work with other disciplines and local people in order to build on opportunities and overcome constraints on farmer water supply and control by examining properties of irrigation management at the three general levels. Careful design of improved local farmer organizations in the interface between main and farm system is an activity strategic to the development of irrigated agriculture. It is hoped that analysis of the concepts, variables, and propositions will be viewed as a step in that direction.

3

Irrigation Experience Transfer: The Social Dimension

Mark W. Lusk

The purpose of this chapter is to summarize fundamental, key ideas about social organization that have been derived from project experience in irrigation development, which may have generalizable validity beyond the case level, and which, therefore, may be useful in planning future water management projects in developing settings.

Initially, two general issues are presented. First, the importance of the social and economic aspects of technology choice is explored. Secondly, the significance of organizational design and farmer participation to project success is examined. Discussion of these two fundamental themes in water management is followed by the presentation of brief capsule ideas about the human dimension of irrigation water management which have been derived from project experience. Most of these *lessons* pertain directly to the importance of farmer participation, irrigation organization, and user control to project success. The lessons and concepts distilled from field experience are illustrated with examples from irrigation systems in Africa, Asia, Latin America, and the United States.

Social Science in Irrigation

The social sciences derive generalizable principles, rules, and constructs that explain the orderly processes of nature. Unlike the physical sciences, the social sciences are limited in their precision because of the variability of human behavior across time, space, culture, and environment. Human beings are among the most complex types of phenomena scientists study, in part, because they behave willfully based-on factors which often cannot be measured through direct physical processes. Because of the wide variability of human behaviors and individuals, social scientists typically study large groups to control for the many differences among individual actors. Thus, the social sciences are fundamentally probabilistic--that is to say they rely on statistical methods to derive general principles about human groups and behaviors in given contexts.

We noted in Chapter 1 that the behavior of human groups in sharing natural resources such as irrigation water, is an area that has been studied relatively little by social scientists. The result of this is that there are few laws, principles or working constructs yet derived that can be applied based on solid statistical evidence. This may frustrate physical scientists who work on interdisciplinary teams when they clearly have relatively solidified theories and models which may accurately depict the behavior of systems: hydrologic systems, crop systems, and so on. A useful analogy for physical scientists to consider in this regard is to think of social science in irrigation much as if it were like meteorology. Like human systems, weather systems behave with little predictability at a local level while only coming into clear focus when observed over long time frames and wide regions. Yet, obviously the weather can have a dramatic effect on the operation and management of irrigation schemes. Likewise, human behavior and organizations appear to be quite unpredictable and erratic when viewed at the local, individual project level, but given the benefit of long time periods and wide geographic spans, human behavior begins to look much more alike than not. Of central importance is the observation that human behavior is the critical factor in how well an irrigation system is managed and, therefore, how productive that system will be.

After decades of project experience, social scientists and other team disciplines have been able to distill certain key guidelines and recommendations that will be useful in guiding social, organizational and management aspects of future projects in irrigation development and on-farm water management. This chapter presents these observations in a condensed format with the following reservation: General rules of system operation and management always require adaptation to local circumstances for such rules to be effective. Given the very wide differences in human culture and behavior across regions, it is a given that such local fine tuning is more critical for sociological findings than for other disciplines.

If science consists of the generalizable knowledge about natural phenomena that can be applied systematically, then we can say that it is *portable* from one context to another with equal validity given local considerations. In irrigation science, the degree of portablility varies considerably by field of study. Engineers contend that their knowledge, being based on predictable physical processes, is almost completely portable. Nonetheless, there are certain *glitches* in soil physics (such as swelling or shrinking soils) that inject an element of unpredictability into the best designed schemes. In addition, many systems are, in fact, very poorly designed or shoddily constructed, with the result that engineering knowledge is not properly applied.

Agronomists also contend with multiple local variations in climate, soils, seeds, and diseases that require local field testing prior to extension to the farmer. If we were to say that agronomy is 70 percent portable and 30 percent site specific in its application to local agriculture development, we would have to follow that social science is oppositely 30 percent portable and 70 percent site specific. Thus, any irrigation development or water management project that seeks to achieve a respectable measure of success must allow for sufficient time to study the local cultural and social organizational system as it pertains to the management, not only of water, but also its bearing on conflict, law, rule-breaking, and other key factors.

Social Factors and Technology Choice

Early agricultural projects by foreign donors in developing nations tended to be designed as if the goal of development was to emulate the achievements of industrial nations in most respects. Factors which were stressed included aggregrate economic expansion, the development of a strong export sector, the emergence of an internal consumer market, and *modernization*. Many of these factors were reconsidered after it became evident that some nations were falling far short of their development goals. For example, the idea of aggregate economic expansion was challenged as such growth did not necessarily accompany an improvement in social welfare, and indeed, problems of equity in the distribution of increased income often had the unintended consequence of contributing to social and political unrest. *Basic needs* strategies emerged as a response. Although basic need strategies are often not carried out because of an overiding interest in security issues, today most economic development projects attempt to balance aggregate or *trickle down* strategies with the basic needs approach (Selowsky, 1979; Vernon & Ruttan, 1990).

Similarly, the effort to expand the export sector by increasing the agricultural production of cash crops was problematic for several reasons. The volatility of international agricultural commodities markets had disastrous results for those countries that relied heavily on one or two exports. The cases of sisal in Tanzania, tea in Sri Lanka, coffee in Brazil illustrate the vulnerability of singular export economies to global price fluctuations. Also, as the agriculture sector in countries modernized in order to produce such commodities at competitive rates, the input costs for production rose significantly at the local level. This had the effect of pushing subsistence farmers into less desirable lands as prime lands were purchased to make way for more competitive, commercial farms. The long-term effect has often been that land tenure became even more inequitable, agriculture inputs were monopolized by rural elites, and countries became less self-sufficient in food production (de Janvry, 1981). Consequently, most agriculture development projects are now cognizant of the need to balance commercial export farming against considerations of land reform, domestic food production, labor effects, and the welfare of the rural subsistence farmer.

In short, some early projects had the effect of increasing the dependency of the economy on foreign inputs, reducing self-sufficiency, and aggravating the plight of the rural poor. As social scientists documented these processes, competing strategies emerged to mitigate the undesirable and unintended effects of poorly-designed development projects. Among them was the strategy of utilizing *appropriate technology* in order to (1) use local materials, (2) encourage self-sufficiency, (3) reduce energy dependency, (4) cut input and import costs, and (5) provide local employment and maintain local expertise.

The ideas that are central to the appropriate technology strategy are worthwhile to consider in any rural development project for the above reasons, yet it would appear that appropriate technology has become confused with low technology. Texts that survey the approach rarely, if ever, note that in selecting a technology for a sustainable development goal, sophisticated and perhaps even capital intensive technologies may be the most cost-effective and socially desirable in a given context. Indeed, many engineers resist the idea of appropriate technology because it has come to mean using the lowest common denominator in technology transfer and seems to deny them the full use of their skills in solving rural problems. The idea of sustainable agriculture faces similar obstacles.

Appropriate Technology Can Be High Technology

In Africa, some microwave transmission towers for rural information or telecommunications systems in arid regions are solar-powered and technically sophisticated. In India, rural education programs are disseminated by television to remote villages using a geostationary communications satellite. Other examples abound. Census workers might compile data on microcomputers for later transmission to the central bureau. Field workers in irrigation schemes use programmable handheld computers to compile data in the field without access to university computing facilities. Brazil's national effort to replace expensive imported gasoline with domestically-produced ethanol is exemplary as it helps the nation reduce imports, stimulates domestic agriculture production (sugar cane), encourages local technology development (alcohol engines), and reduces dependency on volatile international petroleum markets.

Insofar as irrigation technology is concerned, there is no evidence to support the idea that farmers are too ignorant or conservative to adopt and utilize technically-sophisticated technology, given requisite extension support and the infrastructure to maintain and support the technology. As noted in Chapter 2, too often, agriculture development specialists assume that negative characteristics apply to traditional and subsistence farmers because of their lack of education and social isolation. What they fail to recognize is that the farmer is the local expert whose success or failure in farming may spell survival for the family, and that the conservatism with respect to experimenting with new techniques or tools is an adpative, rational utilitarian response to the perilous nature of agriculture.

An example from the highlands of Guatemala may serve to illustrate this concept. The USAID-sponsored Utah State University On-Farm Water Management Project in Guatemala involved the transfer of gravity-fed, on-demand sprinkler irrigation systems to traditional farmers who had used erratic canal irrigation or relied on rain-fed agriculture. These highland farmers quickly adopted and utilized a relatively sophisticated engineering design. Ironically, the greatest resistance to project implementation came from government officials who felt that the technology was too complicated. The project involved the distribution of water to between four and fifty families per pipe. The plastic pipe and the on-farm equipment, such as emitters, had to be purchased, operated, and maintained by the users. Each pipe served approximately 900 hectares. By the end of the project, 40 pipe systems had been installed and were in operation under farmer control. By 1986, an additional 30 pipe systems had been put into place without foreign donor assistance. This was due, in large part, to the clamor of farmers seeking to adopt this technology in their district. The systems were paid for through subsidized credit to the users who were obligated to pay off the loan in nine years. Surprisingly, nine year loans were paid off in two years or less.

Given that on hilly or mountainous fields, sprinkler irrigation is usually the technology of choice because of application efficiency and improved water and soil conservation, the effectiveness of this technology is often preferable to the even more costly alternatives of land levelling or terracing (Keller & Plocher, 1984). Sprinkler technology was suited to the topography, but equally important, it was adapted to local circumstances by virtue of the availability of reasonable credit and extension services, and because farmers were given the private control to run the system in their own best interests. Currently, the same technology is being tested in the Ecuadorian Sierra. It is illustrative to note that a team, comprised of a Guatemalan engineer and two *campesino* technicians (actual project farmers serving as paraprofessionals), is in Ecuador adapting the system to the local context. In designing projects, it is crucial that we be flexible in our choice of technology, involve farmers in the design phase, and not underestimate the capacity of the farmer to utilize sophisticated technology.

Technology Choice

Crucial socio-economic considerations in the choice of technology include: (1) Can the system be maintained? (2) Can it be afforded, and is it profitable? (3) Who owns or controls the technology? (4) Does the technology type fit the local organizational system? (5) Does extension support the technology? and (6) Is the technology fully exploited?

Maintenance

One of the most fundamental questions to be considered is the maintenance of the system. It is not uncommon to survey irrigation systems that are less than ten years old which are already in a state of disrepair and disarray. It frustrates development specialists and government officials in no small measure to see that multimillion dollar projects are functioning at a fraction of their projected efficiency due to overgrown canals, broken pumps, siltation, scouring, clogged pipes, and so on. The initial reaction, of course, is to look to the farmer for an explanation and to find fault. There are several problems in this line of reasoning. Farmers, like other workers, seek to optimize their outcomes by engaging in behaviors that maximize their individual returns. Simply stated, farmers are rational and self-interested individuals like everyone else. This is not to say that at times their behavior does not appear to be irrational to the outsider, but rather that when the incentives to act in any given way are examined from the farmer's point of view, the apparent irrationally of their acts begins to make sense.

In examining the maintenance of a system, we typically see that if the system is collectively owned by a large group or is owned by government, it falls more quickly into disrepair. If, on the other hand, the technology is owned by individuals or by small, closely knit groups that have face-to-face relationships based upon trust, the system is maintained quite well. As discussed in Chapter 1, rational choice theory and exchange theory in sociology, and public choice theory in political science suggest the reasons for this. The maintenance of a large collective asset by individual users is less likely because the incentives for farmers are to *free ride*. This means that it makes more sense to an individual to engage in work that is productive (and visibly profitable) at the farm level, such as field work or cultivation, than to engage in work which benefits the collectivity. It is not because people are irresponsible or because they do not consider the collective group to be important, but that their own limited resources, of which their labor is the main asset, are best utilized from their point of view, in immediately furthering their own more immediate ends. They can diffuse the costs of maintenance, or more aptly of non-maintenance, to the larger group and bear only a fraction of those costs personally. We will see how this logic of collective action also bears on other facets of irrigation, such as farmer organization, but it is clear that it should be considered in choosing a technology that can be maintained with local resources and for which there is an *incentive* to maintain.

The case of Sri Lanka, where irrigation is at least a thousand years old, is illustrative. Major government efforts to intensify agriculture through large-scale irrigation began in the 1960's. These projects have often been highly disruptive of existing, traditional, small-scale irrigation. The effect of state organized expansion and rehabilitation was for farmers to become disengaged from operations and maintenance activities. By perceiving that the tank or canal is the property or responsibility of the state, farmers assumed that even though they were beneficiaries of the new schemes, the responsibility for the system was not their own (disengagement). The new centralized and more intensive irrigation bureaucracies have tended to decrease local involvement by failing to build on existing water user organizations and cultures (Skogerboe, 1986).

Systems cannot be maintained without the requisite infrastructure. One of the best examples of this can be found in the Acion Ble-Dire Project in Mali, West Africa. Efforts were made to introduce multiple small diesel pumps into a remote region of the arid Sahel. The project failed in no small part because there was a chronic shortage of mechanics, fuel, and spare parts. The distance of markets, the cost of inputs, and the costs of transportation were factors as well, but all illustrate the necessity of designing projects which consider the infrastructural supports (Moris & Norman, 1984). One might have expected that the project would have been successful because farmers owned and controlled the technology--a factor that usually contributes to favorable results, but the technology was not adequately supported.

Another example of this theme is what social scientists call orphan technology. This can come about in a couple of ways. When a machine is introduced into an economy and the flow of parts and replacements is interrupted by national trade policies such as import restrictions, the machinery will likely fall into disuse unless it is of such a design that tool and machine shops can manufacture replacements or parts with local materials. It is not uncommon, given the complexity of government bureaucracy, for one national policy within a country to come into conflict with another. For example, the goal of reducing expenditures on foreign goods may contradict the need to increase domestic agricultural production. Similarly, an unrestricted import policy can result in an orphan technology of sorts. If, in a developing nation, distributors are importing pumps of many different brands or types, the lack of standardization can lead to local shortages of parts and/or maintenance expertise. In either case, we see the need to coordinate agricultural policy with economic policy--a fundamental premise of an integrated rural development strategy.

Economy

Obviously, we must also consider whether the system technology is truly affordable and profitable or *window dressing* for political objectives. Prior to design and construction, the economic viability of the project must be detailed carefully and realistically. If the project is to be financed by a foreign or domestic loan, who is going to pay it back: the users, the taxpayers, or some combination, and in what ratio? To expect subsistence or mixed agriculture farmers to finance major projects based on projected increases in cash crop income is highly unrealistic. Front end subsidies or grants are usually required to make major schemes viable as the start up technical and organizational costs are of an order of magnitude incomprehensible to traditional agriculturalists. While it may be possible to offset costs through the production of hydroelectric power for example, major systems require a corresponding level of social organization and financing which is usually beyond the capacity of local user's groups or federations (Freeman, 1986).

In evaluating the success of irrigation development projects, we often neglect to consider the political and social benefits of the program and narrowly focus on the economic rate of return and the increase in agronomic production. By so doing, we forget that governments often have social welfare, employment, political, and security objectives as well. The design of systems, technically and organizationally, is rooted in the social benefits which are accrued in the development and use of the scheme; benefits such as employment, self-sufficiency, use of indigenous materials, rural social welfare, and the enhancement of local expertise.

Irrigation schemes that are heavily reliant on foreign exchange and imported technology are likely to fail due to economic factors alone. The high costs of design and construction often require external capital which is obtained at no small cost in terms of interest, and in relation to alternative development investments that are not funded. In East African irrigation development, costs of US $4,000 per irrigated hectare are not unusual. It is nearly impossible to get a favorable return on this level of investment, even if it is justified in terms of reducing food imports and dependency (McConnen, 1986). In Mauritania, for example, Koita (1986) found that the cost of producing rice domestically was 59 to 124 percent more than imported rice. It may be that these production costs could be reduced over the long-term by increasing crop intensity or multiple crop seasons, but an interesting aspect of African irrigation that is often overlooked which may be far more important in accounting for poor performance is the relative lack of economic freedom. In many African countries such as Mauritania, farmers are not free to choose what they may grow, nor does an open market determine what price they will earn for their crop--the state regulates market prices. Under such a system, it is not surprising that irrigation investments, particularly poor ones, should yield a poor return.

Irrigation Organization

Technology never functions independently of the human energy required to design, utilize, and maintain it. In this sense, technology is a part of culture. If culture consists of all that is transmitted from one generation to the next, then it has two dimensions. Social culture can be defined as consisting of the symbols, ideas, knowledge, norms, and rules by which societies function (software) and physical culture consists of the tools and artifacts which people utilize in their work (hardware). Although quite elementary, we may lose sight of the fundamental interconnectedness of the two.

Irrigated farming is based on physical and social interdependencies between multiple cultivators. Major irrigation development schemes or on-farm water management projects are thus never solely a matter of changing individual farmer behavior in order to improve efficiencies, but also are based on building the connections between farmers and between farmers and organizations.

Accountability and Control

By virtue of having to share the natural resource, water users are inextricably connected by the physical distribution, and by the socio-organizational system that manages that distribution. As we have seen, the technology of irrigation cannot be considered independently of the organizational framework. More importantly, the type of technology can be matched to the organization, or vice versa. Let us consider the simple case of gravity flow, canal-fed, surface irrigation. The control by users over water availability is a function of the number of upstream irrigators on the delivery system, total water supply, net seepage, and evaporation. To have any effect on the delivery of water to the field channel, the agriculturalist has options which correspond to the above factors. The farmer can try to increase total water supply to the system, try to reduce seepage through canal lining or other techniques, or can work with other users to address inequities in downstream delivery--all inherently organizational problems.

To affect total water supply requires that the farmer *reach up* into the system to influence those who control the main works. To line canals, or otherwise reduce water losses necessitates influencing those who control the middle level irrigation organization: the canal company, lateral group, water users associations, *subak*, or what have you. To address inequities in the delivery of water to the tail requires that all of the users on the system cooperate with some method of allocation that is fair and proportional. We can, thus, see that each method of positively affecting on-farm water availability requires a high level of farmer organization because individually a farmer cannot effectuate significant changes in water management, except at the field level. Even at the field level there is little a farmer can do strictly on an individual basis to change the availability of water or its efficient use. This is because most on-farm methods of improved application efficiency (such as leveling, terracing, drip or sprinkler irrigation) are based on significant labor investments beyond the individual's own capacity or technological improvements that require coordination with extension, credit, marketing, and maintenance. Improved performance of irrigation at both the system or farm level is rooted in the organization of water management.

The function of irrigation organization is to design and manage the institutions and physical structures which economically deliver water in a reliable and timely manner with the highest possible degree of control at the farm level. Water has little or no value if it is not of sufficient volume for crop needs, or arrives at the field channel too early or late. Given that crop stress dramatically affects yield, the issue noted in Chapter 2 is to provide the root zone with the amount of water needed to keep the crop above the wilting point while not over irrigating. The volume of water must be predictable throughout the growing season so that users can plant in relation to anticipated supply. Furthermore, the predicted volume must be available when needed.

Fundamental to irrigation organization is the problem of control. Irrigation organizations, from the main system to the ditch group that are not influenced or shaped by the irrigators, will invariably produce inefficiencies (Parlin et al., 1986). This is related to the idea of organizational accountability. The actors in inefficient irrigation schemes usually do not bear the costs of their inefficiencies. If, on the other hand, those who bear the organizational costs also capture the benefits, we might expect inefficiencies to decline. The task of social scientists in working with irrigators, is in large part, devoted to the development of irrigation organizations that link productive farming and water use practices with positive economic outcomes. A key part of this process is to build in mechanisms of accountability; meaning simply that those who must take the risks and pay the costs of farming also capture the benefits. Farmers are willing to make enormous investments of energy, labor, and cash when they perceive that they are in control, their risks of failure are low or moderate, and they will be able to reap the benefits of their work. What we often see in the field, however, is that the primary users of water have little or no control over its administration or delivery. Individuals whose livelihood is not dependent on the efficient and timely delivery of water (unelected bureaucrats) usually have the greatest say over how it is allocated and managed.

The conventional argument in favor of the bureaucratic administration of water is the presumed requirement for *specialists* to manage complex engineering and allocation systems for multiple users. While this is more often the case than not, it does not obviate the logic of farmer control. There are many ways in which the specialized tasks of managers can be merged with user control or influence. If it is possible in the local cultural and legal context to build organizations that privately own the waterworks, and through which users can collectively control the administration of water rights and shares based on water law, then a very efficient method of building in farmer control can be created. Another approach which may be suitable in societies that restrict the private management of natural resources is to democratize the public organization by electing water administrators or commissioners who then supervise the specialists in the agency. The accountability to users in both cases is built in through the participatory process.

Organizations and Technology

For our purposes, the three basic levels of irrigation organization are: farmer associations, command area organizations, and central or main system organizations. Each level of organization has its corresponding roles, responsibilities, constituencies, and overall focus. It is simply unrealistic to think that farmers can be expected to invest a significant part of their energies in the coordination and management of local command area organizations or the immediate concerns of farm management and local water control. The task, therefore, is to match organizations with their corresponding level of activity, and to incorporate accountability at all adminstrative levels to the needs of the primary users.

As noted previously, farmer organization is not a panacea, but rather a method of incorporating an often neglected aspect of sound management practice into the administration of irrigation--the central importance of the consumer. This is central to the functioning of organizations at all levels of water management. One of the several considerations involved in this process is to examine the divisibility of the technology. A technology is described as divisible if it can be utilized by a variety of consumers on an individual or small group basis with little or no group coordination by breaking the technology or tool into functional parcels or elements. Conversely, a non-divisible technology is one which cannot be readily divided into usable units, and is used by large groups of consumers who must carefully coordinate their access to the resource (Freeman & Lowdermilk, 1981).

A simple example of computers can illustrate. A large mainframe computer at a major university is a relatively non-divisible technology. Although it is designed to be used by individuals, no single individual can fully exploit its power. Such a system is beyond the capital resources of individuals or small groups, and indeed, its computing capacity is not appropriate to a small number of users. The tool is inherently suited to multiple uses, a large number of users, and time sharing. Its expense and maintenance requirements are high, but are assessed against all of the users, and thus, are relatively small. One great advantage the tool has is its enormous computing power--a capacity most users will need only infrequently. The mainframe computer can be contrasted with the microcomputer--a relatively divisible technology. The micro has the advantage of being relatively inexpensive (within the means of most American farmers), relatively portable, and it requires little coordination or sharing to exploit its power. As a disadvantage, it is obviously much less powerful than a mainframe, and is restricted in its applications. If an agriculture extension service were seeking to help farmers process data for improved water management, for example, two options might be to collectively utilize a mainframe computer through a time-sharing network, or to provide program and applications assistance to a variety of individual microcomputer users. In either case, the tool selected will significantly affect the organization required to accomplish the task. Irrigation technologies are analogous. Water is used by individual farmers and is highly portable, but the technology of water delivery varies widely in its divisibility, sophistication, capital costs, and management requirements.

Hardware requires corresponding software. Highly divisible technologies such as fertilizer, seeds, fuel, field ditch systems, and low cost farm inputs require relatively little in the way of social cooperation and agreement in order to manage them locally. The layout and management of command area irrigation structures and the cooperative purchase and use of large farm machinery on a share basis exemplify technologies that require a high degree of social organization. Finally, the management of watersheds, the building of reservoirs, dams, or major diversion structures, require very high levels of coordination, capital, and cooperation to effectively manage them (Freeman & Lowdermilk, 1981).

As a project selects a technology type, it is important that the appropriate organizational arrangements be designed. This is a task that must be grounded in a socio-organizational assessment of the local environment. Organizations which are already in place can be used effectively as a springboard for managing the new tools. The social scientist can evaluate existing networks and organizations in this regard. Because local cultural conflicts and social cleavages will aggravate project coordination and implementation efforts, it is useful to assess the project environment for such factors in advance of the technical design phase (Lynch, 1985).

Another crucial element in the design of irrigation organizations is flexibility. Irrigation schemes are not static--they are dynamic and evolving. Structures change physically and must be re-designed and maintained periodically. Water law evolves as does

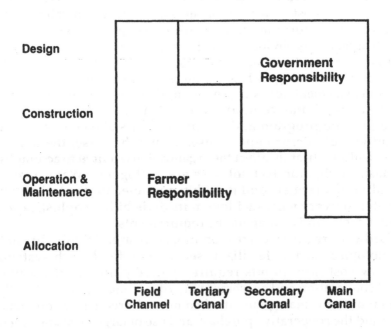

FIGURE 3.1 Private and Public Roles in Large Scale Irrigation.

national and local policy. Markets, soils, budgets, and technology all change. Institutional flexibility allows water users and bureaucrats to correspondingly modify the organization of water use, rules, technology, and law according to the requirements of the system at a given point in time. Such flexibility, however, is based on freedoms that public projects do not typically incorporate as can be seen in projects like the Mahaweli Scheme in Sri Lanka or the Gezhira Project in the Sudan where the farmers are under the influence of statist policies. One can inquire, for instance, whether farmers can freely buy and sell land or water shares? Can they contract labor, work off-farm, lease or sub-let their property? Can they choose what and when to plant? If not, why not? What is usually gained by regimentation in agriculture is economic stagnation or the emergence of informal underground economies because farmers in such circumstances will rationally seek to optimize their profits and production outside of the legitimate system.

Organizational Character

Organizational design can be considered in terms of its public or private character. We tend to think only in terms of public development, and fail to exhaust private options before deciding to publicly finance a project. State expenditures tend to be consumptive, and are usually based on a subsidy because the social desirability of a project is expected to outweigh its economy. Too often, we neglect to consider that a project can, in fact, pay for itself, and by so doing, avoid many of the pitfalls of public control (Adams, 1983).

A strategy in organizational design is to articulate a balance between the disadvantages of public institutional management and failures of the private market. Among the problems of public managment are fragmentation due to overlapping agencies, public monopoly control by subset of users, a non-economic subsidy, and bureaucratic inefficiency due to a lack of accountability to users. Private sector, or market failures, include resource depletion because of a lack of planning and resource mismanagement (pollution, sedimentation, salinization) due to inadequate accountability to the public. Other market problems are private monopoly control and the inequitable distribution of water, land, and income (Lusk & Riley, 1986).

As discussed in Chapter 1, efforts to privatize irrigation can be appropriate in certain situations. For example, the incentive to overutilize irrigation water can be controlled by requiring user fees, and by privatizing lateral, ditch, and user associations. The single best argument in favor of privatization is that it clearly puts the farmer-owner in control. The approach can fail, however, when it is used to have farmers pay for inefficient and unproductive systems, or when a costly system that has a poor rate of return is expected to be financed by subsistence or low income producers. Multiple beneficiary financing (municipal, hydroelectric, etc.) can distribute the burden of costs in many settings. On the other hand, it is not unreasonable to require farmers to pay a portion of their increased earnings resulting from intensive irrigated farming to cover the costs of managing and delivering irrigation water. Indeed, having to pay such costs on a fee per volume basis is an important conservation incentive.

Certain tasks best *fit* a public representative organization: the regulation of monopolies, the control of water quality, watershed management, major construction for multiple beneficiaries, and the equitable resolution of social conflicts within a constitutional framework. These advantages must be weighed against the problem of bureaucratic accountability wherein the constituency is neglected by public workers who have few incentives to increase on-farm water supply and efficiency.

In examining the extent of farmer control, it is necessary only to determine at what point the management of water is turned over to the farmers. If the water is not under farmer control until it reaches the fieldgate, then the system is essentially a centralized one. If, on the other hand, farmer control or influence reaches far up the management system, state control is moderated. State involvement may be indicated in major systems or conveyances--the management of a 300 mile long canal in Pakistan for instance. On the other hand, the public sector can better handle its *appropriate* tasks, such as the adjudication of water rights, when it is not burdened with *ditchriding* chores and local conflict management.

Scale

Some countries are turning away from the development of capital intensive, major, high technology, large-scale schemes in reaction to the poor performance of these systems in economic terms. Early analyses of the failures of large-scale schemes emphasized political culture as an explanatory variable. Wittfogel (1957) argued that such major centralized irrigation systems emerged and were successful only in despotic hydraulic cultures or in nations which have strong, bureaucratic, centralized, or colonial forms of government. The state has played a crucial role in this type of irrigation development, and it has been argued that the absence of such centralized power limits the success of such schemes. It is more likely that the performance of the larger systems has been limited in part by the marginal quality of the lands available for new development and, ironically, by the despotic nature of these systems. Thus, in Asia and Africa there is a trend toward investment in smaller-scale, community operated systems which emphasize rehabilitation and improved on-farm water management (Barker, Coward, Levine, & Small, 1984; Ssennyonga, 1986).

This recent direction in new investment is based on the premises that community-based irrigation schemes entail lower construction and operation costs, promise a better return, enhance farmer involvement and initiative, and are less disruptive of existing irrigation organizations. Evidence from Peru (Lynch, 1988) and Kenya (Ssennyonga, 1986; Gichuki, 1985) supports these premises.

As these shifts in irrigation strategy occur, it will be useful to consider a wider range of organizational alternatives to centralized bureaucracy. Because hierarchical central bureaucracies tend toward depersonalization, corruption, regimentation, and coercive authority, the search for effective irrigation organizational designs must be flexible and open to the utilization of existing rural networks. Rural segmental networks that can be utilized to advantage may include: kinship groups, clans, tribes, religious groups, voluntary associations, and cooperatives (Barclay, 1986). The advantages of such informal, segmental groups is due to the effectiveness of group sanctions against rule-breakers, the group allegiance that can be invoked for labor, the higher degree of individual freedom afforded, and the economic interdependencies of members that act as buffers to individual hardship.

Lessons Learned in Irrigation Development

The danger of compiling key lessons that have emerged from irrigation development projects is that historically, rural development has been characterized by simple solutions to complex problems. To propose certain ideas as lessons derived from project experience is to risk oversimplification. Recognizing that local site specific considerations must always be primary in project design and management, general issues can be identified to guide future and ongoing water management projects. The brief and non-exhaustive list here is derived from the project experiences of technicians at Cornell University, Colorado State University, and Utah State University, and is based on project reports, interviews, and workshops.

(1) *The farmer cannot take all the risks.* Traditional and subsistence farmers tend to be conservative in accepting or adopting new technologies, ideas, or farm practices. This is not because they are *ignorant and illiterate* as some would have it, but because traditional and peasant agriculture tends to function on a zero sum basis. The costs of risking new techniques are sometimes great when compared to the benefits of increased agronomic production. The risks or costs may include poor harvests, increased input expenses which cut into profit, changes in land tenure, changes in household structure or economics, and rural dependency. While a project may increase farm income and presumably rural welfare, changes in the rural economy may also have the effect of aggravating inequities in income or land tenure, destabilizing rural social organization, and creating rural elites. Traditional farmers tend to be *risk-aversive* because in marginal subsistence agriculture it is economically rational behavior. Projects can reduce farmer risks by thorough, local field testing and a gradualist plan of implementation.

(2) *Time heals.* Irrigation organizations and schemes that are older tend to work better. This is, in part, due to the phenomenon of *cultural lag*--a process by which social organization cannot keep pace with technological change. Technology invariably has social repercussions. It is slower to work out the social implications of technology than it is to develop new technology.

It appears that irrigation and water management projects have a long life cycle and several stages of evolution. Initially, there is concern with the material technology of a system and the management of hydrologic matters. A later stage is concerned with increased agronomic production. The final stage is centered around the organization of water rights, water use, system maintenance, and social conflict. The life cycle of an irrigation scheme is much greater than the time frame of a development project. Project goals are often designed to be achieved in as little as 5 years, when some have estimated that a new irrigation system takes at least 50 years to stabilize. The two immediate implications of this are that project goals should be stated more modestly, and when possible, existing irrigation organizations must be capitalized upon in order to reduce cultural or organizational lag.

Another implication concerns project flexibility. Because of the evolutionary or dyanmic character of irrigation, change must be anticipated, and allowance for change and conflict needs to be built into the organizational design. How can this be incorporated? Flexibility implies allowing cultivators to grow what they deem to be economically feasible. It means budgets that plan for maintenance, replacement, re-design, and expansion. Institutional flexibility means allowing water users and bureaucrats to change the organization of water use, water law, and rules according to changing environmental and market conditions. It anticipates the need for prompt and equitable resolution of user conflicts and deviance. Given that farmer income may paradoxically fall when productivity rises due to market saturation, the question arises as to the flexibility and responsiveness of the external system to allow for new marketing strategies, export potential, and subsidy policy.

(3) *Bottom-up can be belly up.* Because of the serious problems associated with top-down centralized planning and control, a reaction of social scientists and technicians has been to advocate the development, expansion, or rehabilitation of small-scale community based systems. This new direction is an important one, and holds considerable promise for activating local capacity to operate and manage irrigation. Such programs would be at great risk, however, if they fail to recognize that the organizational costs for small-scale development are not necessarily less than those of large-scale schemes, unless existing organizations are utilized and strengthened. The central idea driving the logic of promoting community based systems is that they tend to be more efficient, and are responsive to farmer needs. Care is needed, however, not to use local organizations as scapegoats for project failure, nor to dump technically sophisticated systems onto traditional users.

Bottom-up development may require front-end capital subsidies to start up a project, expanded time frames to accommodate farmer skepticism, and long-term agriculture extension to transfer and adapt new technologies. Any project that requires water user fee increases must involve farmers in planning so that such fees are economically viable and are supported by users. Furthermore, it must be recognized that agriculturalists have little incentive, given the demands of local farm management, to be heavily involved in command area or main system management. It may be more appropriate for them to employ staff for these tasks that represent their interests. In the Intermountain West of the United States, for example, most private irrigation companies are farmer-owned and employ paid staff to operate the command area or district. Such private arrangements ensure that the irrigator is served at the middle level of organization while free to manage water at the field level.

Rather than thinking of *bottom-up* as the preferred method of water development, it may be more useful to consider that *hardware requires corresponding software.* This would mean that instead of working from the top down, or vice versa, organizations are created to serve specific technological and management ends based upon their scale, constituency, and budget.

(4) *Agriculture extension or bust.* Projects that make shortcuts in agricultural extension services are taking great risks in the probability of a successful outcome. Not only is extension crucial to the adoption of the technical innovations, but is also central to the socio-organizational aspects of project performance.

In the farmer participation pilot projects of the Philippine National Irrigation Administration (NIA), community organizers (CO's) were used with considerable success in developing grass-roots organizations capable to participate effectively in decision-making, planning, securing water rights, construction, maintenance, and cost control. In contrast with the usual practice of hiring technically trained bureaucrats to extend technologies, the CO model recruited local educated workers who were trained as catalysts to organize farmers and act as interfaces between the agriculturalists and the NIA. Of interest also, is that engineers, agronomists, and other technical project staff were also trained and *bureaucratically re-oriented* with respect to farmer participation and the coordination of technical activities. This bi-directional participatory approach to extension resulted in improved maintenance, greater farmer acceptance of their financial obligations, and a reduction in water user deviance, such as theft and ditch sabotage (Bagadion, 1985).

The CO model developed in the Philippines has influenced other projects, most notably the Gal Oya Scheme in Sri Lanka. In the Gal Oya, Institutional Organizers (IOs) were recruited to live in the villages with farmers and organize farmers for institutional development. In conjunction with local leaders who were selected to be Farmer Representatives (FRs), a cadre of extensionists and indigenous farmer leaders were able to activate local initiative for improved water management (Uphoff, 1986b). Projects in Nepal and Peru have demonstrated the usefulness of intensive front end agriculture extension. What has yet to be carefully considered are the organizational costs of this extension activity--is it cost beneficial to invest heavily in the software (organization and extension) of project design? While no definitive answer is available, it is clear that to date the vast majority of project funds have gone to hardware (technological) aspects of projects, often at the expense of effectively organizing the management of that technology (Uphoff, 1986a).

(5) *Use the carrot and the stick.* The starting point for the under-standing project success in relation to farmer productivity and water management is with the incentives and disincentives (sanctions) which shape farmer behavior.

As the Farming Systems Research and Development perspec-tive has demonstrated, the logical beginning point for understanding the productivity and organization of agricultural development is at the farm level. By looking at water use, technology, markets, etc. from the farmer's perspective, greater insight can be obtained about development problems. Basic to the socio-organizational analysis of farming systems and farm water management is the examination of the incentives that motivate the producer. Projects and irrigation organizations that give this only tangential con-sideration will not effectively manage conflict, deviance, fee payment, or conservation.

Project managers may attempt to use sanctions to control wa-ter use (fines, water cutoff, etc.) without balancing them against incentives. In this type of organization, one can anticipate that farmers will continually disengage or retreat from the system. They will assume a passive approach to maintenance or fee payments as the organization is driven by their retreat to monitor them more and more, in effect, paternalism.

Conversely, the organization of water may center around in-centives. Rules are designed which reward the desired pro-so-cial, utilitarian behavior. If the organization seeks timely pay-ment of water user fees, for example, it also provides the irriga-tor with the correct volume of water at the time the farmer needs it for crop demands. The organization must also assess water fees that are economically consistent with the market value of water--fees that reflect the capacity of irrigation to boost agronomic production, but which do not also require the agricultural user to subsidize other water development objectives (hydroelectric, navigation, fisheries, municipal). An approach which is built on incentives will be much more effective in motivating local capacity to man-age water, yet using incentives alone is not typically enough to achieve optimum organization.

In irrigation organization, incentives need to be balanced against sanctions. To collectively utilize a resource requires consider-able order and discipline based on rules that promote the group good without sacrificing individual rights--the delicate task of water law. This can be illustrated in relation to typical problems of the common pool.

A common pool resource such as irrigation water, is one which is held collectively by a community of users, all of whom have rights of access to it. The logic of the commons is that each individual will attempt to draw on the resource to maximize individual benefits, but since ownership of the resource is collective or public, the costs of such utilization will be borne by the group. This can lead to resource depletion as individuals rationally seek to use the water to their advantage, yet do not bear the contingent costs. The phenomenon of the *free rider*, an endemic one in irrigation, is the result of the failure to carefully balance sanctions and incentives. In the absence of institutional restraints to abuse, the resource will be exhausted or monopolized by a powerful few. This logic helps explain related problems, such as deforestation, overgrazing, and water pollution (Lusk & Riley, 1986).

To counter the tendency to deplete or abuse the resource requires institutional arrangements (rules, regulations, laws, fees, property rights, and markets) which provide both opportunities as well as constraints in the public and private sectors (Easter & Welsch, 1986; Coward, 1986b). User deviance, such as fee non-payment, water theft, and corruption, will disappear only when it is promptly and equitably sanctioned (disincentives) and when there is profit in user compliance (incentives).

As discussed in the first chapter, this logic can be implemented in part by: (1) opening markets to ensure profitability; (2) requiring reasonable user fees to limit overuse; (3) keeping water user associations small to encourage better surveillance and the formation of face-to-face relationships and communities; (4) electing water judges from the users to manage conflict according to collectively designed rules and regulations; (5) privatizing lateral, ditch, and user associations to maintain autonomy and incentives; and (6) building in bureaucratic accountability by command area and main system managers to the water user constituency through the election of commissioners and other key functionaries.

As with any operational *rules of thumb*, these general recommendations may be applied only after adaptation to site-specific factors in the local environment if they are relevant to the conditions of the system at all. Locally, costs may be cultural factors such as *loss of face* rather than monetary currency, and incentives may be *gaining status in the community* rather than, or in addition to, farm profit. An ethnomethodological grounding of system and organizational design will ensure that incentives and sanctions are built into the project which are rooted in the specific cultural and economic environment of local farming. It is thus useful to search for aspects of resource management which have general application as long as we consider that rules of organizational design must always be tailored to the local socio-cultural milieux.

(6) *Plan for unintended effects.* It is not a contradiction to plan for the unplanned. Development projects (planned social change) will often have effects on the social and physical environment which are not a part of the project objectives; indeed the effects are, at times, contrary to project objectives. If a project design is adaptive to the dynamic and evolutionary character of irrigation, then it will more ably respond to the challenge of unanticipated effects.

The most significant of the unintended social effects of irrigation can occur when the target group is not the same as the impacted group. For example, a new scheme may displace a population (while ostensibly benefitting another) and, in the worst case, make little or no allowance for the affected group (Moris & Thom, 1985). The Srisailam Hydroelectric Scheme in India, for example, flooded 107,000 acres of farmland which had supported 100,000 people, causing enormous hardship to local villagers who had to be relocated. Many such major water development projects contribute to the already serious problem of rural to urban migration (Goldsmith & Hildyard, 1984, 1986). In other cases, projects that are designed to improve rural welfare may paradoxically aggravate land tenure maldistribution, encourage *landlordism*, and strengthen rural elites--unintended socio-economic effects (Parlin & Lusk, 1988).

Similarly, traditional water user claims may be overlooked or revoked by uninformed planners. Small plot farmers may be pushed off the land as the economy of scale changes under irrigation. Irrigators who rely on *tail water* or drainage from upstream irrigators may find their ditches dry when conveyance and application efficiencies are improved nearer the head. Diesel pumps and tubewells may increase water supplies for some, while drying up the hand dug wells of the poor. The list of unintended effects is as long as the history of planned social action (Boudon, 1982). Therefore, it is crucial to consider who are the actual beneficiaries (corporations, domestic elites, local farmers) and who must bear any ill effects--planned or not.

The Senegal River Development Project exemplifies an environmental effect. Although stabilizing water supply for irrigation and hydro-electricity, the Senegal River dams and others along the estuaries of West Africa will have a major effect on the fishery. One estimate is that the net food gain resulting from irrigation will be offset by the food loss from the fisheries (Riley, 1986). It is not uncommon for planners to overlook the *intangible* economic value of soil preservation, water replenishment, wildlife, tourist appeal, and climatic stability (Goldsmith & Hildyard, 1986).

Dams and flumes may produce negative health effects in environments where diseases have a water-based vector. In fast moving waters of West Africa, *onchocerchiasis* (river blindness) may increase near flumes or spillways. Likewise, near the stagnant waters behind dikes and dams, the risk of malaria and schistosomiasis (*bilharzia*) rises (Moris, 1985). Such undesirable effects must be weighed against expected gains associated with increased production, such as better health, income, nutrition, and improved household welfare. An alternative approach to this problem is an integrated rural development strategy which simultaneously considers correllary developments in pest control, water quality, and sanitation.

In other cases, the best planning just does not come up with the desired result. The Plan Piloto Irrigation Project in Peru, for example, used a disease-free Dutch seed to demonstrate the effect of irrigation frequency and fertilizer on potato yield under varying irrigation schedules. Although the project was a success in terms of showing extensionists how yields could be dramatically increased, the potato produced was too large to be successfully marketed given local tastes (James, 1986).

The key is to coordinate the multiple effects of social action when possible, and to be flexible for the challenge of the truly unintended consequence when it occurs. Too few projects are genuinely multi-disciplinary, and as a result, rural development has often not proceeded in an integrated way. Irrigation water development cuts through fields of professional activity so a successful project will simultaneously consider seed, fertilizer, pesticides, technology, research, extension, marketing, transportation, processing, health, production incentives, credit, land, tenure, and other factors. Integrated rural development for irrigated agriculture is inherently a multi-disciplinary endeavor.

(7) *Fit the technology to the cultural landscape.* Social cleavages within a community will be manifested as irrigation conflict when these divisions occur within a single distribution system. An excellent way to undermine project cooperation is to overlay a physical structure onto a heterogenous community of users. Social divisions that can be of critical importance are social class, ethnicity, religion, and political party factionalism (Lynch, 1985, Uphoff, 1986b).

With respect to social class, projects can be endangered if support is primarily from the elite rather than based on a community-wide consensus. If land, wealth, and power are highly concentrated, the benefits of irrigation will be very uneven and community support will be correspondingly low. For whatever reason, if the effects of a project are to accentuate the maldistribution of assets, the propensity of the scheme to be operated and maintained with local resources are greatly reduced.

Irrigation development is a strategy to increase production from existing or newly-opened lands, and is typically directed toward benefitting the rural poor. The vast majority of this group are living on marginal lands at a subsistence level of production, often with tenuous claim to the property. It is their marginal economic status and low level of production which has historically been the *factor* limiting social differentiation. Uneven economic development disrupts the level social strata, and thus, has the potential of producing new social tensions and aggravating old ones.

Subsistence farmers could benefit greatly from increased production yet, ironically, projects that increase output may determine the welfare of the target group. This occurs, in part, because of the increase in the value of the irrigated land. The incentives for absentee landlordism, land invasions, corruption, and illegal land appropriations grow in proportion to the land's value. An increase in production through irrigation is also accompanied by correspondingly greater input costs in labor and capital which, in turn, raises the peasant farmer's level of risk and vulnerability (Raynolds, 1985).

Existing area rivalries or divisions will likely be heightened if distinct cultural, ethnic, religious, or political groups have to cooperate in managing a basic agricultural input. Such tensions will become more evident during periods of scarcity. Irrigation committees and associations tend to be strongest in homogenous, egalitarian communities (Lynch, 1985).

The Gal Oya Project in Sri Lanka illustrates the difficulty in managing water in a heterogenous social environment. Despite intensive work to organize farmers into a cohesive unity, the head of Gal Oya is predominantly Sinhalese (Sri Lanka's dominant majority) and the tail is mostly Tamil and Muslim (minorities). The physical distribution along the system reflects the social hierarchy, and there has been no mechanism in the irrigation organizations to solve the problems of minority farmers (Parlin & Lusk, 1988; Zolezzi, 1985).

If divisions of this nature exist, then a legal basis must be included in project design to exclude political or social factors from affecting allocation or distribution. A mechanism for rule enforcement and conflict management should be incorporated which guarantees the rights of minority users. A better approach yet (when it is cost feasible) is to engineer *around* the social landscape so that groups function within their own technological domain.

The foregoing capsules can be represented as several working axes or parameters. These can, in turn, be referenced against propositions from other fields to form a holistic representation of a given irrigation scheme. The result of multi-disciplinary opportunity analysis of a project will be a plot of system position on an evolutionary path with respect to key parameters and considerations. This representation yields an understanding of overlapping problems and suggest opportunities for change and adjustment (Keller, 1988).

(1) *Private vs. public.* Is the scheme a state-owned and developed project, farmer-owned and operated, or some combination of the two? Public projects may require investment in decentralization and farmer organization to build in incentives. Private projects may require investment in infrastructure, extension, and land reform.

(2) *Water price.* Is the water free to the user, very costly, or in between? High priced water encourages conservation, but lowers productivity by raising input costs. Free or very low cost water encourages overuse, reduces the incentives to cooperate and participate in irrigation organizations, and lowers system productivity due to over-irrigation. We propose that price is related to productivity and participation. An optimization point can be found between free and costly water, dependent upon the local economy and culture.

(3) *Water scarcity.* Is water abundant per user hectare? The incentive to cooperate and to conserve is reduced in systems which have abundant water resources. On the other hand, the propensity for social conflict is greater in water scarce systems.

(4) *Extension intensivity.* Is the project fully supported by extension, or is there little or no technical/educational coverage? Low support projects are successful only when technical transfer is minimal. Significant innovations require technical and social science extension to assure adoption and to mitigate against unintended effects.

(5) *Costs.* Capital, maintenance, and organizational costs per developed hectare are a fundamental consideration in the viability of a project given local markets. The payout must be reasonable for farmers based on the local economy, or farmers will disengage.

(6) *Farmer type.* Are the farmers refugees, or in-migrants? The time horizon and extension requirement for projects with relocated farmers is greater than for stable agriculturalists who are adopting innovations. A related concern is: Are the farmers experienced with irrigation (of whatever technology), or are they used to rainfed agriculture? The case of relocated farmers from rainfed systems is an especially difficult one as was observed in the Mahaweli Scheme of Sri Lanka.

(7) *Subsistence or surplus farming.* Is the target farming system based on a traditional subsistence, mixed, or cash crop economy? Subsistence farmers tend toward risk aversion and resist innovation. The social effects of technical innovations are more strongly felt in homogenous, egalitarian, traditional farming systems, and must be planned for and mitigated. In cash crop systems, marketing and rural dependency are cardinal considerations.

(8) *Scale.* Is the system small-scale and possibly locally controlled? Is it large-scale and organizationally complex? The scale will determine the type of organizational pattern which is optimal. Considerations in this regard are: participation, representation, bureaucratization, privatization, multiple ownership, divisibility of technology, and organizational costs. Large-scale projects can encourage local participation when they are private, representative, segmental (decentralized), and equitable.

(9) *Age.* Older systems tend to be more stable and free of conflict. This does not mean, however, that they are necessarily more productive, efficient, or equitable.

(10) *Cultural diversity/homogeneity.* Is the social environment relatively heterogeneous? Systems which cut across class, political boundaries, or sub-cultures tend toward greater social conflict, deviance, and disorganization.

(11) *Conflict and deviance management.* Are there institutional mechanisms in place for the management of conflict between users and bureaucracies, between interest groups? Conflict is a normal sociological phenomenon present in all cultures and societies, and, therefore, must be anticipated and managed equitably. Likewise, deviance can be expected as users will test the system even in optimal schemes. Is rule breaking promptly, publicly, and equitably sanctioned?

(12) *Water and land property rights.* Do farmers have secure property rights that correspond to their responsibilities? Many irrigation development projects seek to quickly transfer to the users the responsibility for maintenance and operations costs at the tertiary level. In the absence of secure property rights to their land and/

or water, farmers will assume that these responsibilities belong to the state. Few cultivators will seek to protect an institution's investment when their own rights are ambiguous or unprotected.

Summary

There are other lessons or capsule ideas which can be derived from project experience. Our purpose here is not to exhaust the list, but to underline the central importance of a few key ideas which have been repeatedly supported by project experience in a variety of cultural settings.

Two general themes have been addressed which can be used to help understand the social dimensions of irrigation water management: (1) the importance of technology choice to project success in a specific physical, economic, and cultural environment; and (2) the centrality of irrigation organizations to the management of conflict, maintenance, water allocation, and system operation.

Finally, with respect to the social organization of irrigated agriculture, it is not possible to overemphasize the importance of maintaining incentives. When incentives are used as a fundamental unit in the social analysis of irrigation, seemingly incomprehensible problems, such as sabotage begin to make sense. It is important that incentives and their counterpart, sanctions, be utilized in addressing bureaucratic as well as farmer behavior. But, like any conceptual tool, incentives cannot be considered inflexibly without local ethnographic adaptation. It may be that the avoidance of conflict or *saving face* may be of importance in one context while reduction of labor inputs is important in another. Using general principles from organization theory, it is possible to adapt development strategies to the local socio-cultural context by focusing on what motivates individuals to induce and adopt rural changes-- incentives.

Our working assumption is that farmers seek to control their resources in order to maximize agricultural production and can effectively do so when they own their land, have a reliable supply of water, appropriate technologies, access to inputs at fair prices, and an open market for their commodities. Integrated irrigation development pursues these ends.

PART TWO

Farmer Participation and Irrigation Management: Case Studies

4

Farmer Involvement in **Water** Management: The Case of Sri Lanka

Bradley W. Parlin

Introduction

Sri Lanka, formerly Ceylon, is an island off the southeastern tip of the Indian subcontinent, bordered by the Indian Ocean on the west and the Bay of Bengal on the east. The 1981 Census puts the population at approximately 15 million, roughly 25 percent of whom live in urban centers and 75 percent in rural agricultural areas. The population growth rate of 1.8, while low by Asian standards, is putting increasing pressure on the nation's food security as the potential for expanding arable lands progressively diminishes with the development of large-scale rural colonization schemes based on irrigated agriculture. The tropical climate of Sri Lanka reflects little seasonal variation and the island is clearly demarcated into a wet zone (south) and dry zone (north) which is a result of the varying influence of the southwest and northeast monsoons. The impact of the two monsoon periods results in two distinct cultivation periods, the *maha* (wet season) from October to December and the *yala* (dry season), which occurs from April to June.

Farmer Participation in Water Management

The central purpose of this paper is to explore the different forms of farmer participation in water management activities found in Sri Lanka today. The wide range of variation in types of irrigation schemes combined with both ancient and modern forms of wate𝚛 management provides a rich source of information useful in illustrating the potential range of farmer involvement in irrigation activities. The social organizational contexts of farmer involvement in traditional minor tank water management will be compared to modern large-scale irrigation schemes based upon colonization.

Historical Background

Physical evidence of massive ancient irrigation structures are scattered throughout the dry zone. In fact, many of the tanks (*wewas*) utilized in the modern large-scale schemes of contemporary Sri Lanka date back to the pre-Christian era. One of the largest (six thousand acres) tanks in northeastern Sri Lanka, Parakrama Samudra near the ancient capital of Pollannaruwa, was built under the guidance of King Parakramababu in the 11th Century. Even in this early period, the king encouraged water conservation demanding that, "Not one drop of water should run into the sea without being used," (Abeyratne, Brewer, Ganewatte, & Uphoff, 1984). However, successive invasions from southern India, internal warfare, pestilence, drought, and famine, apparently led to a decline and abandonment of these reservoirs from the 12th Century onward (Gunasekare, 1982; Jayawardene, 1984).

The British Colonists exiled the last Singalese king in 1815, and thereafter pursued an aggressive policy of developing export crops (tea, rubber, coconuts), while for the most part, ignoring the development of indigenous peasant agriculture. This long-standing practice resulted in the stagnation of domestic agriculture and the growth of food security deficits at the time of independence. A country, which in its early history, exported food grains due to its hydraulic and agronomic expertise, became under the British, dependent upon the substantial importation of paddy (rice) with a concurrent loss of scarce foreign exchange (Ponnambalam, 1980; Roberts, 1980). Today, one can still see remnants of the old civilization in the form of *purana* (ancient) villages in the dry zone. A central feature of these old villages is the reservoir, where water from the monsoonal rains is stored and used in paddy cultivation directly below the tank. These small-scale tank cultivation systems are supplemented by both rainfed highland cultivation (*chena*), in which the jungle is slashed and burned with seed sown during the rainy season, and *gangoda*, or home gardening. The social organization of these small-scale tank schemes, as we shall see, stands in sharp contrast to the large modern irrigation conveyance systems based on colonization and resettlement (Farmer, 1957; Jayawardene, 1990; Uphoff, Wickramasinghe, & Wijayaratna, 1990).

*The Social Organizational Context of Minor Tank Water
Management*

Minor tank schemes comprise a substantial component of irrigation activity in Sri Lanka today. There are approximately 30 thousand
minor tanks, 52 percent of which are being used in active cultivation activity. The remaining 48 percent of minor tanks are abandoned,
but it is estimated that about 7,000 of these can be brought into
productive use and could sustain about 500 thousand people (Medagama,
1984; Begum, 1985). Minor tank schemes are said to contribute
approximately 20 percent of annual national paddy production
in Sri Lanka (Begum, 1985).

The role of water management in indigenous peasant agriculture
in Sri Lanka is the domain of the village irrigation headman (*vel
vidhane, diyabalana, wattei vidhane*) who typically manages a tract
(*kandam*) of 300 to 500 acres. The irrigation headman appoints assistants
(*adigaries*) and all are remunerated (*suthanthiram*) by cultivators
in proportion to the amount of land under cultivation. Typically,
remuneration consists of 1/4 bushel of paddy for 2 and 1/2 acres
(ARTI, 1985; Leach, 1961). The irrigation headman is elected by
cultivators. While the tenure of the irrigation headman has varied with colonial and post-independence regulations from one to
three years, evidence suggests that they have typically come from
the land-owning class (*goigama*) with high status in the feudal hierarchy
of the village and frequently have virtual lifetime appointments
(ARTI, 1985).

The functions of the irrigation headman focus upon the management of irrigation water (supply, distribution), organizing the
cleaning and maintenance of conveyance channels and check works,
enforcing fencing obligations, resolving cultivation related conflicts, including issues of land tenure and property rights, the
organization of the cultivation (*kana*) meetings, and so forth. The
major formal sanctions used by the irrigation headman to ensure
compliance and cooperation of cultivators to water management
rules are to report violators to the village council, and the with-
holding of water issues.

There can be no question, however, that informal mechanisms
of social control provide the most salient source of the headman's
power. The *purana* villagers are, by and large, skilled agriculturalists, and village social structures have long-standing cooperative traditions such as *attam* (shared labor) *shramadana* (labor donated
to the community) and *bethma* (water sharing arrangements during periods of drought). Cooperation among *purana* villagers is

not a matter of choice, but rather necessity. Kinship ties, marriage ties, market relations, caste relations, and intimate, primary, day-to-day relationships based on history and biography become a powerful binding source of cohesion and conformity to village rules, policies, and practices in all dimensions of life. Both physical and social survival become intimately dependent upon docility and conformity to the group's expectations. Thus, the irrigation headman's control over water management is reinforced by coercive, formal and informal norms of village life. In fact, many cultivators confer a higher status to the role of irrigation headman than to the village headman or council chief to whom he is formally subservient (ARTI, 1985). In summary, minor tank water management under the *vel vidhane* system is a bottom-up approach in which cultivators elect the irrigation headman in a social milieu in which cooperation between both headman and cultivators is based on traditions of mutual interest and family and community intimacy.

The Social Organizational Context of Modern Large-Scale Irrigation Schemes

From the early 1930's, the government of Sri Lanka started colonization schemes based on irrigated agriculture. Abandoned tanks were rehabilitated and landless peasants from the wet zone were given land below the tanks for paddy cultivation. Necessary irrigation infrastructure was developed and water from major river systems was diverted to augment supplies for both *maha* and *yala* cultivations, since there was almost always insufficient water to carry out a risk-free *yala* cultivation. While over 110 major irrigation schemes are in operation today in Sri Lanka, this chapter focuses on two of the largest systems by way of contrasting the different roles of cultivator participation in water management in small versus large systems.

The Mahaweli Development Scheme

The major river systems of Sri Lanka originate in the central mountains and radiate in several directions into the sea and are the major source of water utilized in the emerging resettlement schemes. The most important river system in this regard is the Mahaweli *Ganga* (river) which flows from its origin in the central highlands, northeast through major portions of the dry zone, to its confluence with the Bay of Bengal at Trincomalee. It is the Mahaweli which serves as the major resource of the Mahaweli Development Scheme which is one of the largest and most ambitious irrigated colonization programs in the world today. The evolution of the Mahaweli Development Scheme had multiple goals, all potentially contributing in some way to dramatically impacting the quality of life of rural and urban poor. The restoration of dry lands, coupled with the intensive introduction of high-tech agricultural techniques (double cropping, improved rice strains, intensive cultivation and programs of water management efficiency) would hopefully move Sri Lanka towards food self-sufficiency. While many *purana* villagers would be displaced by the new impoundments and canal systems, thousands of landless peasants from the densely populated wet zone would be selected to land ownership as colonists in the emerging system. The construction of head works, canals, and hydroelectric units would provide much needed off-farm employment, and make a quantum leap in maximizing energy self-sufficiency. All of these goals would have positive implications for political stability as well (Hameed, 1977; Jayawardene, 1990; Herring, 1985).

The master plan of the Mahaweli Development Project provided for the development of 900 thousand acres of land, of which 654 thousand acres were new lands, and 246 thousand acres of existing land were to be provided with supplementary irrigation. In the first program of Mahaweli Development (System H), 24 thousand families were settled.

Approach to Water Management

Those selected to be colonists in the Mahaweli contrast sharply with the *purana* villagers of the minor tank schemes. Selectees often had little understanding of dry zone survival skills and cultivation practices. In addition, unlike the homogeneous traditional villagers, the heterogeneous selectees did not have the safety net of an established set of cooperative normative structures based on kin relations to rely upon. Among the many problems experienced in the Mahaweli colonies were: high rates of crop failure, hidden tenancy, credit default, and inefficient water management (Nott, 1985; Scudder, 1985). It would appear that many of these problems can, in substantial part, be traced to the top-down approach to water management utilized by the Mahaweli administration and planners. An irrigation assistant for each unit (250 farmers) was appointed by the system administrators to organize turnout groups, oversee water allocation and distribution, canal and check works maintenance and conflict resolution; roles formerly assumed by the irrigation headman. The irrigation assistant would then request that farmers select two representatives for their turnout group which typically consists of 10-20 farmers. One farmer representative would deal with agriculture extension matters, and the other with water management issues. Both farmer representatives were to be trained by the Mahaweli agency in special classes on agronomy and water management techniques and policies, which they would, in turn, disseminate to turnout group members.

Research findings suggest that the top-down approach to water management utilized by the Mahaweli agency has been fraught with difficulties. One study of System H found that only 26 percent of farmers sampled reported membership in a turnout group (Alwis, 1983). Additionally, farmers tended to view their turnout group representatives as often insensitive extensions of the Mahaweli bureaucracy, who all too often used their training to serve their own interests and to exploit turnout group members (Jayawardene, 1984). Research findings of the USAID-sponsored mid-term evaluation of the Mahaweli scheme suggest that cultivators frequently complain that they seldom see the irrigation assistant, and that he often preferred to give orders rather than solicit the ideas and suggestions of water users themselves (Scudder, 1985; Nott, 1985). This, in turn, has led to the progressive alienation and disengagement of peasant cultivators from the Mahaweli bureaucracy and provided a nurturing environment for widespread deviant irrigation practices such as water theft, the destruction of control mechanisms, and subsequent conflict between head- and tail-enders in the system (Parlin, Jayawardene, & Amarasena, 1985). As Nott, (1985:107) observes, "Some settlers were of the opinion that their participation through the turnout groups, farmer leaders meeting, etc., was requested merely to meet the formal requirements of the quorum, not to hear the settler's opinions and requests."

In sum, the top-down approach to water management initiated by the Mahaweli bureaucracy can best be described as pseudo participation which has led to ineffectual involvement on the part of cultivators, resentment, hostility, and the emergence of what Scudder (1985) describes as, a culture of dependence.

The Gal Oya Scheme

The Gal Oya colonization scheme is located in southeastern Sri Lanka and was initiated at the time of independence in 1948. The system serves 30 thousand cultivator families and irrigates 120 thousand acres. From its inception, the scheme was beset by problems and inefficiencies, and the physical system gradually deteriorated. Encroachment had increased the area under irrigation substantially, thereby contributing to persistent water shortages and conflicts between original colonists and encroachers who lived off the system inefficiencies, water theft and return flow (Abeyratne et al., 1984; Uphoff, 1985; Zolezzi, 1985; Ganewatte, 1985). The government of Sri Lanka (with funds and technical assistance from USAID) decided, in 1978, to rehabilitate the left bank of the Gal Oya system. Along with improvements in the physical system, it was decided to involve farmers in the design and subsequent stages of rehabilitation development. Cornell University staff, working closely with the Agrarian Research and Training Institute (ARTI) in Colombo, were to develop a program of farmer organization that would, in contrast to the Mahaweli approach, be bottom-up in nature.

Encouraged by the success of farmer involvement programs in the Philippines, a self-corrective *learning process* approach, as developed by Korten (1980) was initiated. The project involved the utilization of institutional organizers (IOs) to live with and organize turnout groups of 10 to 15 farmers at the field channel level. Thus, farmers were to be direct participants in the planning, design, and construction phases of the project (Uphoff et al., 1990).

The role of the IO was to act as a catalyst to facilitate farmer involvement in all phases of the rehabilitation activity as both teacher and learner. Importantly, the IO would not dictate how farmers were to be involved in decision-making, but rather facilitate group decision-making among the farmers themselves as to the form, function, and extent of their involvement in the rehabilitation process (Wickramasinghe, 1982). The IOs recruited were recent university graduates and included social scientists and extensionists who were familiar with paddy cultivation, had experience in the dry zone, participated in village level social organizations, and were willing to undergo the rigors of living in remote rural areas. All IOs were trained by ARTI personnel in problem-solving related to water management and interpersonal dynamics of group decision-making (Wickramasinghe, 1982). The first IOs were fielded in 1981, and while the program was officially terminated in 1985, the government of Sri Lanka has funded the continued involvement of IOs for the next few years. Initial efforts were plagued with turnover amongst IOs and problems in establishing a nurturing relationship with the irrigation bureaucracy. The results of this bottom-up approach to farmer participation appear, nonetheless, to be encouraging. Increases in yields, more efficient water distribution, and reduction of farmer conflicts exemplify gains claimed by the project coordinator and others who have studied the project (Uphoff, 1985; Wijayaratne, 1984). A good deal of the success of the IO program has no doubt been the ability of the project administration to identify a normative basis upon which resource mobilization could evolve. Most of the colonists were Buddhist who were familiar with the religious concept of *shramadana* (labor donated to the community) which is a traditional method for priests to mobilize community resources for temple building and maintenance. The IOs were able to convince farmers that the maintenance of canals and diversion structures could be based on *shramadana*, and reports suggest that this approach to operation and maintenance at the turnout level was very successful (Uphoff, 1985). In short, a religious tradition was essentially secularized in its application to water management activities in the project area.

While the positive results of the Gal Oya farmer participation experiment parallel similar reports from a variety of third world settings (Philippines, Malaysia, Nepal), one must be cautious in interpreting reports and research findings. There can be no doubt that participation holds many potential benefits, such as increases in productivity, improved water management, conflict resolutions and more efficient resource mobilization. However, the virtual absence of sound systematic empirical research data on the Gal Oya project makes accurate assessments of the farmer participation program impossible. Subjective, impressionistic, case level trip reports and research utilizing rapid rural appraisal techniques are too often devoid of sufficiently sophisticated data bases to arrive at solid conclusions. A confident assessment of the Gal Oya project must be based on quality, longitudinal and/or comparative empirical studies which, to date, do not exist. The true test of the program will be its sustainability; can the reported spirited fervor of farmer involvement in water management in Gal Oya survive the eventual withdrawal of USAID or government of Sri Lanka support? This remains an important question mark with respect to future prospects of farmer organization in Sri Lanka. As Wickramasinghe (1982:4) observes:

> The record of such rural participation and local organization attempts has been a generally poor one, especially in terms of sustained successes. Frequently, initial results have been promising, but over the longer run, rural organizations usually have collapsed once external stimuli have been withdrawn.

Summary and Conclusions

As we have seen, the social organization of water management in traditional minor tank systems in Sri Lanka stands in sharp contrast to those of the modern large-scale systems of the Mahaweli and Gal Oya projects. In small tank schemes, farmer involvement in water management is bottom-up with the *vel vidhane* exercising control of water supply, distribution, operation and maintenance, and conflict resolution through binding rules of cooperation based on village tradition, kinship ties, market relations, and ethnic and religious homogeneity.

The massive Mahaweli scheme, based upon colonization and thereby a heterogeneous settler population, lacked the traditions of cooperation, kinship ties, institutional infrastructure and social networks that contribute to farmer involvement in water-related matters. In short, anomie (normlessness) characterizes settlement communities. In order to encourage settler participation, a top-down approach (external bureaucratic) was initiated and the available information suggests that the results have included farmer apathy, lack of commitment to turnout groups, alienation, resentment, deviant irrigation practices, hidden tenancy, and poor on-farm water management.

In the case of the Gal Oya left bank rehabilitation program, a bottom-up system of farmer involvement in water management was initiated through the use of institutional organizers. While both the Mahaweli and the Gal Oya schemes are based upon resettlement, important differences characterize the two systems. In Gal Oya, the colonization was mature since the original settlement began in 1948. While there was little cultivator participation at the outset of the rehabilitation activities, the settlers lived in well-established hamlets characterized by primary social arrangements which provided a nurturing environment for potential farmer involvement through the IOs as change agents. This allowed the IOs to effectively tie settler involvement to the tradition of *shramadana* in Gal Oya, a feat which would be next to impossible in the heterogeneous communities of strangers of the Mahaweli.

5

Irrigation as a Privleged Solution in African Development

*Jon R. Moris**

Introduction

Some years ago, Albert Hirschman coined the term *privileged problem* for problematic situations which may have been present for a long time, but which seem to require an organized solution by means of public intervention (Hirschman, 1963). The problem of African drought is an excellent example. Especially at a time when Europe and America hold huge amounts of food rotting in storage, it is no longer thought tolerable that hundreds of thousands of African peasants should die from drought-induced famines. But, *recognition of a privileged* problem can also stimulate adoption of privileged solutions, material and organizational technologies which seem self-evidently suited for dealing with problem needs. A privileged solution is not thought to require testing and modification. The answer will seem to lie at hand, and what matters is simply to find the resources and will to act. In Africa, irrigation projects have often enjoyed a privileged status among some policy-makers. They seem the obvious solution for modernizing production, minimizing food imports, removing food deficits, and ameliorating the impact of drought. We shall suggest that this explains why African governments continue to invest in modern irrigation, despite its high costs and poor performance within sub-Saharan Africa. There are many lessons one can learn from the Af-

*The material in this chapter was first published by Moris, J. (1987), in *The Development Policy Review*, vol. 5, pp. 99-124.

rican irrigation experience. Perhaps the most important warn of the dangers encountered when expensive technologies are uncritically implemented in contexts radically different from those in which they originated. These lessons have an obvious bearing on larger issues, such as whether or not the Asian *green revolution* can be duplicated in Africa and the extent to which technological modernity can, by itself, revolutionize African agriculture.

What is Privileged?

The concept of *privileged solution* employed here refers to preferred investments, chosen because they are thought to answer some pressing policy problem. When accorded privileged status, technologies are more likely to be adopted by reference to the seriousness of the problem than on the basis of their own likely performance under realistic field conditions. We shall argue that in Africa, irrigation is often seen as the universal answer to drought, and thereby escapes detailed justification and local adaptation.

It is not immediately clear from aggregate statistics that modern irrigation has received priority within Africa's development. In Africa as a whole, there are only 6 million hectares (ha) of modern irrigation, mostly under major government schemes, and of these, 82 percent are said to lie in North Africa and the Sudan (FAO, 1986a). In the remainder of black Africa, the bulk of irrigated production occurs under traditional, small-scale technologies. If we include both the traditional and modern types, perhaps 2.5 percent of the cultivable land in sub-Saharan Africa is under irrigation at present (FAO, 1986b). Before we turn to review the performance of modern irrigation in several countries where its adoption has been emphasized--notably in Sudan, Mali, Senegal and, to a lesser extent, Kenya--let us first look at the production setting into which modern irrigation is introduced.

Basically, to irrigate means to employ technology in order to augment and control the supply of water for crops. Whether the application is by diking within polders, or by canals and furrows, flood basins, or pipes (for sprinkler and drip irrigation), the technology demands substantial on- and off-farm investment. As an additional component in the production process, irrigation is almost always more expensive than rain-fed cultivation: in African countries, sometimes *much* more expensive. Of course, in really dry environments (such as in Africa north of Khartoum), irrigation is the only method of growing crops, but even then one should examine the economics of importing food instead of growing it at very high cost.

The case for regarding modern irrigation as a *privileged* technology in sub-Saharan Africa rests upon several interlocking observations. The most basic is that in poor countries which have the option of rain-fed cultivation (a sector being denied the investment it needs), it is hard to justify investment in expensive modern irrigation. Irrigation is also a heavy consumer of water in a generally dry continent where water is frequently scarce: Africa is on average the driest of the world's continents (World Resources Institute, 1986). A single household watering its garden in porous, sandy soils with a hose, may use as much water as an entire settlement without irrigation. Advisers from developed nations sometimes forget that in a semi-arid environment, water itself is the basic resource, subject to an array of competing uses.

Because of porosity, sandy soils are not suited to surface irrigation (except under expensive sprinkler systems). The lands where large-scale irrigation has been developed in Africa consist of alluvial soils with a high clay content near the larger rivers. Such areas are already a critical production resource, supporting not only a large livestock population during the dry season, but also sometimes valuable inland fisheries (Maltby, 1985). Under such uses, the *size* of the area taken over for irrigation may be a poor indicator of its strategic importance under alternative uses. Places like the Lake Chad Plains in Nigeria, the Butana Plain in Sudan, the Awash Valley in Ethiopia, or the lower Tana in Kenya (all having been taken over for modern irrigation) were key dry season forage reserves for traditional livestock producers.

Since the 1950's, many countries have experienced an intensification of farming in the same areas being sought for large-scale schemes, but based instead upon the spontaneous evolution of local irrigation technologies (Richards, 1986). In Nigeria, for example, such farming is known as *fadama* cultivation and depends upon the use of low-lift pumps and shallow wells to support garden crops (vegetables, onions, and spices) grown in the dry season. The Nigerian Government has put a substantial investment into unproductive large schemes, but FAO estimates that these produce only about 10 percent of Nigeria's rice output, and there are 800,000 hectares under informal irrigation, which until recently received no official assistance at all. Even in Madagascar, Africa's largest rice producer, less than one third comes from the organized irrigation sector (FAO, 1986c).

Typically, irrigation planners have insisted that there was no productive use for the land and water required by modern schemes. Twenty years ago this claim was advanced when Kenya's Mwea

Irrigation Settlement wanted to expand into occupied lands along its southern border. In reality, these lands were being intensively farmed by private operators who employed landless households as the labor force and who grew Indian dry-land legumes more suited to famine relief than the wet-rice produced by the scheme (Chambers & Moris, 1973). More recently, in northern Nigeria, the Kano scheme was installed in the upper reaches of the Hadejia River which flows north-eastward into Lake Chad. The water held back by the Tiga dam (and diverted thereby into the Kano scheme) had served a floodplain described as "the most important rice and cotton producing area of Kano State" supporting about 250,000 people (Stock, 1978). In fact, on several schemes such as Ahero in western Kenya or Bakolori in northern Nigeria, the numbers of people evicted to permit scheme development exceeded those subsequently given plots (Baum & Migot-Adholla, 1982). The common feature in all these cases was that farmers were employing simple, indigenously-evolved technologies.

Those promoting modern irrigation view it as a logical and *necessary* alternative to what they suppose is inefficient, high-risk traditional cultivation. It is taken for granted that farmers, given the chance to participate in modern irrigation, will wish to do so, and further, will accept the external financing, bureaucratic controls, and guided production regime this entails. Even the crops to be grown are usually stipulated. Farmers entering such systems discover they are expected to become specialized, perhaps mono-crop producers, farming according to scheme or agency dictates in settings sometimes far removed from markets and support services.

The building of a dam and its associated canal network usually results in the creation of some sort of surrogate organization to take the initiative and to coordinate farmers' individual involvement, i.e. what is termed in the literature as *scheme*. Here the significant fact is the absence of medium- and large-scale farmers throughout much of sub-Saharan Africa. Small-holders with only two or three hectares do not constitute an attractive market individually for the commercial supply of modern irrigation technologies.

This explains why, although there has been a spontaneous diffusion of small Japanese pumps, one does not find farmer-led adoption of perimeter irrigation which depends upon large pumps or dams and requires a joint distribution network. The overhead sprinkler irrigation which a single operator in Nebraska or Australia might install, requires cooperation from several hundred small-holders in a typical African context. Thus, implementation of modern irrigation in Africa goes hand in hand with the imposition of larger units, either commercial estates or parastatal schemes, to act on farmers' behalf. In neither instance do small-holders make the choice of technology themselves; the choice is bureaucracy-led rather than demand-led.

Unfortunately, experts who are called in to do project design (whether under private or public auspices) frequently fail to appreciate the local factors which strongly influence the economic success of irrigated farming. The few genuinely successful projects have been large-scale sugar estates, usually under private management, where there is minimal small-holder involvement and the whole complex constitutes an enclave of modern technology unconnected to its surrounding economic and administrative environment. The Hippo Valley and Triangle Estates in Zimbabwe's lower Sabi drainage constitute one such complex. Otherwise, when schemes are meant to incorporate small-holders, there have been many examples of poor design and inappropriate technology choice. The engineers who have been responsible will sometimes admit this has been the case.

None of the water experts contacted could point to any large agricultural projects built over the past 10 years in black Africa that have made a dent in the current famine. "Africa has borne the brunt of a lot of bad advice, bad engineering and bad intentions and they are paying the price for it now," says a U.S. engineer who is the area manager for Africa at a major international water resources consulting firm (*Engineering News*, 1985:10).

On larger projects, design tasks get parcelled out between various agencies and consultants. Often, the initial site identification is done by UN or AID agency experts, perhaps considering only the physical factors. Then a succession of other consultants work out details and do the economic appraisal, usually in cooperation with a local irrigation parastatal such as Niger's ONAHA or Kenya's National Irrigation Board (NIB). This means that it is already accepted that irrigation is the preferred option. Eventually, a management team will be recruited, but by then the important choices will have been made. External consultants are usually based in Europe or America, and return to base after each phase in project development. This institutional setting inhibits learning from experience at the local level. By the time feedback from farmers can occur, the physical investments are locked in, and, in any event, the experts in charge of irrigation design will have departed.

The Bura West Case

Kenya's Bura West project provides a good illustration of how difficult it can be to establish modern irrigation in remote African environments. (Except as otherwise indicated, all references here to the Bura West scheme are derived from Gitonga's [1985] FAO case study.) As early as 1948, the colonial government proposed two 40,000 ha irrigation schemes for the lower Tana basin, but detailed surveys showed them to be uneconomic because of poor soils and the remoteness of the area. In 1956-57, the colonial government went ahead with the Hola Pilot Irrigation Scheme, about 600 ha, utilizing detainee labor. An FAO/UNDP survey in the early 1960's revived the idea of large-scale development with its proposal for a rehabilitation of Hola and the formation of a Tana River Irrigation Authority to manage development of between 100,000 and 120,000 ha of potentially irrigable land (the size reckoned necessary to achieve a 10 percent rate of return).

The site for the Bura settlement consists of riverine lands on the lower Tana, located in an isolated, semi-arid zone about 80 km downstream of Garissa (the nearest town, itself quite remote from the rest of Kenya). The area had supported about 15,000 Pokomo farmers and some 6,000 Orma semi-nomadic pastoralists. There was virtually no infrastructure. Initial planning suggested a scheme of 14,000 ha divided into two segments, one on either bank of the Tana. After further surveys, World Bank support was gained to begin the first phase (Bura West) under a US $98.4 m (Kshs 766m) project involving six donors and the Kenya Government with the east side not yet thoroughly studied. The 1975 feasibility study undertaken by Dutch consultants was redone at Bank insistence by British consultants in 1976. Meanwhile, financial arrangements were finalized to permit construction to begin in late 1978 for 6,700 ha on the west bank. The objectives for the scheme were to settle landless families, to increase production of food and cotton, to save foreign exchange, and to reclaim under-utilized, semi-arid land (Ruigu, Alila, & Chitere, 1984). The economic rate of return expected was 13 percent. (Key World Bank documents include the appraisal report [No. 1446-KE] issued in 1977 and the midterm evaluation issued in 1985.) The first settlers arrived from up-country and began moving onto the land in 1981.

Costs escalated because of design changes recommended by successive consultants (65 percent in real terms by April, 1979), and alterations in the size of the scheme made either as a result of findings during the preparatory studies, or because of financial constraints. One funder (the Commonwealth Development Corporation) withdrew, leaving the Kenya Government to make up the difference. There was pressure from other funders and the government to get work started before all the data had been analyzed. The cheap weir initially proposed for the west bank only was changed by the consultants to a bigger structure farther upstream that could irrigate both banks. This, in turn, necessitated a long supply canal with expensive inverted siphons at each point where it was crossed by seasonal drainage. For the first phase of the west bank, only the consultants proposed a pumping station, arguing that this was the best temporary option pending completion of studies on the larger structure and the larger area. Faced with unfinanced additional costs, and an unfavorable soils report on the east bank, the Kenya Government was forced, in 1983, to reduce the west bank area to 3,900 ha even though some of the infrastructure to serve the larger area was already in place.

The social, agronomic and economic aspects have been equally problematic. Health facilities were built three years late so that there were *many fatalities* among the first tenants from the virulent strains of malaria found in Kenya's sub-coastal environment. A survey conducted by a non-governmental organization found 52 percent of the project's children were malnourished and 81 percent underweight. For the first year that tenants were on the scheme, there was no primary schooling at all, and many tenants left. Housing cost Kshs. 32,000 per unit by 1983, "raising serious doubts whether the tenant farmer will ever be able to pay for his house" (Gitonga, 1985:30). The planned fuel-wood plantations, so vital to relieve women from walking many miles to seek firewood, were not even planted by the time the tenants arrived (Vainio-Mattila, 1985). Cotton yields during the first three years (ranging between 2,000 and 2,200 kg/ha) were about half of potential yields, in part because of pumping breakdowns, but also because of delays of a month or two in land preparation (both being management responsibilities). The pumping difficulties occurred because of shortages of fuel and spare parts, while land preparation suffered because private contractors were unwilling to work in such a remote setting. A University of Nairobi evaluation in 1984 conducted by Ruigu et al. (1984), estimated that 45 percent of tenants received incomes inadequate to meet their basic needs. By 1985, the NIB was incurring an annual operating deficit of about Kshs 45m at Bura. Meanwhile, the Cotton Lint and Seed Marketing Board owed tenants Kshs 15m, an amount the NIB was forced to make up from other sources. Finally, the Kenya Government stepped in and took Bura out from under the NIB's control. However, the Kenya Government continues to seek financing for yet another large irrigation scheme (17,000 ha) on the Tana delta.

From a wider perspective, the problems evidenced at Bura are far more common than official sources will admit. There are other recent African irrigation projects with similar histories, e.g., Nigeria's

Bakolori and South Chad schemes, or Somalia's Afgoi-Mordile Project. What all have shared is the assumption that a site that was attractive from an engineering standpoint was the main requisite for a successful project design. As the Bura case illustrates, knowing the *solution* in advance made it unnecessary to learn about local soil conditions, existing resource utilization, optimal crop combinations, or farmers' interests. The commitment to irrigation occurred before local costs and impacts could be evaluated, and continues *despite* adverse experience. Governments have assumed that modern irrigation must be the technological answer permitting crop production in a high-risk, marginal environment. Instead, actual experience indicates that the quasi-industrial character of the technology makes it particularly vulnerable to disruption under these circumstances.

Where Privileged?

A recent FAO review documents wide variations in the degree to which African countries employ irrigation. In the whole of sub-Saharan Africa (excluding South Africa), Sudan's 1,700,000 hectares under modern irrigation are over ten times the size of any other country's; second comes Madagascar (160,000 ha), followed by Zimbabwe (127,000 ha), and Mali (100,000 ha). These are the only four countries with more than 100,000 hectares under modern irrigation. If one also includes areas used under partial water control and the traditional, small-scale systems, then Madagascar's total jumps to 960,000 ha, followed by Nigeria (850,000 ha), Mali (160,000 ha), Tanzania (140,000 ha) and Senegal (100,000 ha) (FAO, 1986c). Such figures are mere guesses, but they do substantiate our earlier observation that a relatively small proportion of Africa's potentially irrigable land has been put under irrigation. Outsiders familiar with the much higher degree of irrigation development in parts of Asia might argue--as did the papers for a major FAO conference on African irrigation held in April, 1986 at Lomé, Togo--that Africa needs *more* rather than less development of modern irrigation.

Let us start with Sudan, whose irrigation sector is much larger than anything seen elsewhere in sub-Saharan Africa. The first of Sudan's major schemes was at Gezira, begun in the early 1920's, and drawing its water from the Sennar dam. Gezira has been expanded several times and, today, the combined Gezira-Managil scheme totals some 865,000 ha in a single huge block located between the two branches of the Nile south of Khartoum. By the early 1970's, it constituted 12 percent of the total cultivated area of the Sudan, and produced 75 percent of Sudan's long staple cotton (the mainstay of Sudan's export earnings) (Barnett, 1977:6). Long viewed as a success story by the Sudanese Government, Gezira became the model for the New Halfa Scheme (163,000 ha) in the mid-1960's and the Rahad Scheme (126,000 ha) a decade later (Gaitskell, 1959). (Figures on scheme sizes vary greatly in different accounts, but are taken here from FAO, 1986c: 108.) In addition, the Gezira system was copied on various smaller pump schemes along the Nile, later taken over by the government (Rangeley, 1985:16). Thus, despite the huge size of Sudan's irrigated agricultural sector, the institutional framework, based on a Ministry of Irrigation which constructs the water supply and production parastatals which perform basic tillage operations on the behalf of numerous tenants, is the same throughout the system.

The flat clay plains of central Sudan do constitute an ideal situation for large-scale irrigation development. The topography is unusually uniform. The soils, while alkaline, are moderately fertile and retain water; gravity-fed distribution is usually feasible; and the whole zone is crossed by rivers draining higher rainfall areas to the south and east. Existing population densities away from the rivers were low. With the evolution of Khartoum into a major urban center, some form of irrigation development was inevitable.

Even so, the technological system adopted can be viewed as being *privileged* in comparison with what farmers would have chosen on their own. The various production parastatals established by the Sudanese Government to operate each of the schemes paid little attention to farmers' interests. Before cotton prices were raised in 1982, Gezira's cotton, for example, provided the lowest return per man-day (less than half what tenants would get from either wheat or groundnuts) of the four main crops permitted at Gezira, and the highest consumption of water (Faki, 1982). Chronic indebtedness among the poorer tenants has emerged on Sudan's newer schemes, and families still farm under an annual production license despite having been in residence for a decade or more. The description by the anthropologist Sorbo (1985:40) of life on the New Halfa scheme applies broadly to the whole Sudanese irrigation sector, which was, in the early 1980's, seeking World Bank assistance to rehabilitate its *successful* schemes:

Production was low; absenteeism was high; there were repeated shortages of water, vehicles and fuel; pests and weeds invaded the fields . . . required inputs of seeds, fertilizers and pesticides rarely arrived on time; poor storage facilities caused deterioration and losses; and tenant incomes were low and extremely irregular . . .

As in the Gezira Scheme . . . a sense of helplessness pervaded the Scheme and the loss of morale and deteriorating operating conditions were mutually reinforcing.

For the most part, West Africa's modern irrigation development has been located within or adjacent to the Sahelian zone, a transverse band of land lying just south of the Sahara and receiving between 100 and 600 mm of rainfall annually. The two main rivers, the Niger and Senegal, draw their water from the Fouta Djalon highlands in Guinea, but cross into the drier lands of the Sahel zone on their way to the Atlantic. The Niger flows in a huge, 1,700 km loop that brings it to the very margins of the Sahara at Tombouctou in Mali before it bends southeastwards and receives a second infusion of water in Nigeria. As with the Nile in East Africa, so also with the Senegal and Niger, the rainfall received in the highlands at the source is transformed into an annual flood which travels slowly downstream. Farmers along both rivers practice an indigenous form of *flood recession* irrigation by planting crops in succession down the riverbanks as each season's floodwaters recede.

The Sahelian landscape, as just described, provides moderately rich alluvial soils in close proximity to major rivers. Mali gives us the West African equivalent to Sudan: a large country divided sharply into an arid north and a wetter south, and bisected lengthwise by the Niger River with its own, huge inland delta at the heart of the country. The British example at Gezira was not lost upon the French, who in 1932 established a parastatal corporation, the *Office du Niger*, with the grand aim of putting nearly a million hectares of the delta under irrigated cotton and rice production (Fresson, Amselle, Bagayoko, Benhamou, Leullier, & Ruf, 1985). Actual achievements have been about one-twentieth of the initial plans: a dam was built across the Niger in the 1940's, and by 1964 the Office had almost 50,000 ha under irrigation, with a population of 33,000, producing 43,000 tons of paddy and 9,000 tons of seed-cotton (De Wilde, McLoughlin, Gunard, Scudder, & Maubouché, 1967). For over forty years (1935-80), rice yields remained stagnant, varying between 1.4 and 1.9 tons/ha, while a bloated bureaucracy (over 4,000 permanent and 5,000 temporary employees) outnumbered the tenants (5,700 in 1984). Today, 31 percent of the area developed has been abondoned; many tenants are deeply indebted; and over one third (36 percent) were due in 1983 to receive nothing for their season's work after deductions for the Office's services (Fresson, et al., 1985).

Nevertheless, the notion that irrigation could produce an *island of prosperity* for Mali has persisted. In the 1960's, Mali's Rural

Engineering Service brought nearly 100,000 hectares into polder irrigation, spending 38 percent more than its allocated budget -- an effort taken over by the socialist Keita regime from French colonial plans, but one which ironically "was an economic, political, and technical failure which destroyed Mali's plans for rice self-sufficiency as well as its dreams for rural socialism" (Bingen, 1985:36).

The projects continue through different regimes and successive incarnations. The Traore regime in the early 1970's, established Operation Riz, covering some 12,000 ha of rice polders scattered at various points throughout the inland delta, developed with assistance from the French Fund for Aid and Cooperation (FAC) and the European Development Fund (EDF). Rice yields remained, however, in the range of 1,200-1,500 kg/ha. In 1972, the Mali Government split the polders into Operation Riz-Seqou, (financed by the EDF) and Operation Riz-Mopti, developed with World Bank financing. Both employ partial control wherein protective dikes along the river permit floating rice to be grown under a controlled flood after the rice has been seeded within the polders under rain-fed conditions. Between 1972 and 1978, Operation Riz-Segou was expanded from 17,000 to 35,000 ha, with Riz-Mopti encompassing a further 26,000 ha.

The indigenous system within the inland delta employed African rice (*Oryza glaberrima*) and gave yields of between 500 and 700 kg/ha. It was intended that through better water control and substitution of Asian rice (*Oryze sativa*), this would be raised to 1.5 tons/ha, roughly what peasants in southern Mali achieve from rain-fed rice cultivation with no official assistance (McIntire, 1981). Irrigation planners expected that partial control, diked systems would fail one year in ten (because of either insufficient rain to get the floating rice started or a late arrival of the Niger's flood). The failure of the rains twice in succession during 1983 and 1984 had a catastrophic impact on yields from these polders. Nonetheless, the World Bank still portrays Riz-Mopti as a success, with an economic rate of return of 17 percent (FAO, 1986c). However, by the Bank's own figures, mean yields have been only 926 kg/ha, and a detailed case study of the companion Riz-Segou project documents relatively poor performance (Bingen, 1985). Thus, to date, neither the Office du Niger nor the large floating rice polders have realized the Mali Government's hopes, despite several decades of effort and the involvement of multiple donors.

The Senegal River too, has been the focus of successive attempts to introduce modern irrigation as a replacement for indigenous, flood-recession technologies. Traditionally, rice production came from swamp rice grown in The Gambia and Casamance Rivers to the south; farmers along the Senegal River grew short-cycle millets, cowpeas, and flood recession sorghum (Craven & Tuluy, 1981). During French colonial times, however, development efforts were directed at establishing modern, large-scale irrigation on the Senegal delta. The main complication was the annual salt-water intrusion up the river after its flood abated, and also salinity in the underlying sediments. The large-scale Richard Toll perimeter begun during World War II was, after much difficulty, brought into production on a much reduced scale, but required an annual subsidy of between CFAF 8 m and CFAF 50 m every year from 1946 to 1960 (USDA, 1980). The post-independence projects such as Nianga in Senegal's middle valley, were also intended as mechanized, medium-sized polders, but encountered so many technical problems that the government finally abandoned large-scale units in favor of its own, locally evolved approach to small-scale perimeters managed largely by farmers themselves.

The parastatal charged with supervising small-scale perimeters along the Senegal River (SAED) was, at first, heavy-handed and bureaucratic (notable in its dealings with the USAID-assisted Bakel perimeters upstream near the border with Mali) (see Adams [1981], Miller [1985], and Keller, Meyer, Peterson, Weaver, & Wheelock, [1982]). Since then, it has gradually learned how to work with farmers' communities in developing a unique, float-mounted pumping system suited to small perimeters. As of 1983, it was supervising 19,000 ha, mostly under rice, and involving some 35,000 plot-holders (Bloch, 1986). Despite problems, the small scheme element probably provides Africa's most successful officially-sponsored small-holder irrigation with mean rice yields of between 3.5 and 6 tons per hectare.

Even here, the privileged status of large-scaled approaches continues to be evident. During the same period, the Senegalese Government was joining Mali and Mauritania to establish a large river basin authority (the OMVS), charged with developing the whole Senegal basin. Under the OMVS, two dams are nearing completion, one at Diama at the mouth of the Senegal to stop salt water intrusion and the other up-stream in Mali at Manantali to create a substantial reservoir. These large projects have as their aim stabilizing the flow of the Senegal River (thereby eliminating flood recession cultivation) and encouraging navigational use, while providing power generation at Manantali. The fact that Senegal already has an under-utilized railway connecting it with Mali weakens the case for the navigational benefits; the OMVS has, therefore, emphasized the prospects for additional medium-and large-scale irrigation of the very types which have repeatedly failed within Senegal in the past. Senegal's valuable fishing industry will also be affected, to a negative, but as yet, unknown degree.

Africa's most populous country, Nigeria, missed the initial stage of colonial sponsorship for large-scale irrigation. Nigeria (like its neighboring states to the west) straddles a range of ecological zones, from dry near-Sahelian conditions in the north to humid forests in the south. Drought in the north during the 1970's was accompanied by higher rainfall to the south. It is thus, technically feasible for Nigeria to make up any shortfall in northern food output by increased production elsewhere.

Nevertheless, Nigeria's 1962-68 National Development Plan identified the lack of irrigation as one of the critical factors limiting agricultural output. The following account of Nigeria's irrigated wheat production draws heavily upon Andrae and Beckman (1985), chapters 6-8. FAO was asked to prepare an overall assessment, while at the same time both FAO and USAID were doing feasibility studies in the north for the Sokoto-Rima valley, the Hadejia-Jama'are river basin and the Lake Chad basin. The Commonwealth Development Corporation also became involved in investigating the prospects for irrigated wheat production. The resulting reports in the late 1960's stressed the technical feasibility of irrigation, while leaving unexamined its economic cost-effectiveness. This was probably because investments on such a scale then seemed to be beyond Nigeria's financial and technical resources for many years to come.

Nigeria's oil boom in the mid-1970s coincided with the Sahelian drought and rule by northerners. A federal deficit of half a billion *Naira* in 1970 was turned into a N 2 billion surplus in 1974, and by 1977 oil revenues had reached N 6 billion. Three large projects identified in earlier studies and proposed for gradual development through pilot projects were now pushed ahead: the Kano River Project (20,000 ha) 60 km south of Kano and drawing water from the Tiga dam; the Bakalori (or Talata Mafara) Project on the Sokoto River, planned for 23,000 ha; and the South Chad Irrigation Project (22,000 ha) in the extreme north-east on the shores of Lake Chad. In 1975, a separate Ministry of Water Resources was carved out of the Federal Ministry of Agriculture, and eleven river basin authorities were put under it. The share of agricultural investment targeted towards irrigation in the third plan was 30 percent, subsequently raised in 1977 to nearly 50 percent.

To justify irrigation on lands already growing rain-fed crops, the Nigerian authorities proposed a dramatic increase in dry season wheat growing. A study group suggested in 1971 that Nigeria's production (6,000 tons in 1970) should rise to 150,000 tons in 1980 and 300,000 in 1985. Andrae and Beckman (1985) found that less than a third of the Kano Project's planned area was under irrigation and of this, only half was planted to wheat. At most, 5,000 tons were produced, as against project forecasts of 30,000 to 45,000 tons. The Bakalori Project, which was supposed to have reached 20,000 tons in 1982-83, in fact reached only a tenth of that in 1981-82, and even less in 1982-83. Meanwhile, the river basin authority responsible for Bakalori continued to project an output of 32,500 tons of wheat for 1985, a hundredfold increase over the actual 1983 performance. The even larger South Chad Project was designed to produce over 50,000 tons in 1981-82; its output was probably less than a tenth of this projection.

These three projects represent an investment of approximately N 1.2 billion in 1982 prices, with development costs ranging between N 12,000 and N 20,000 per hectare. (In 1982, the official exchange rate put the *Naira* at roughly US $1.5). That this level of support is unwarranted by performance is plainly evident. At Bakalori, for example, the large number of farmers dispossessed from their fields during construction and the delays in paying compensation, brought about a peasant insurrection, which was quelled by military action in which only seventeen farmers are admitted to have been killed. Andrae and Beckman (1985:115-116) suggest that the real number killed may have been many times this total. Another source which deals with the sad Bakalori experience is Adams (1984). This same project suffers from insufficient water ever to irrigate its designed area. The sprinkler system was dimensioned for 24-hour operation, but farmers have refused to irrigate at night, and the high winds often experienced during the day render sprinklers unusable. When wheat is grown, it overlaps farmers' guinea corn crops and threatens household food security. Mechanized operations instituted at great cost at Bakalori and South Chad have been extremely inefficient. High-capacity harvesters are kept stationary, being hand-fed by farmers who must first cut their wheat with sickles. Farmers were paying between N 25 and N 38 per ha water charges, between one-fifth and one-tenth of the actual costs. Andrae and Beckman (1985) calculate that, even if no capital repayment is attempted, the factory gate cost per ton of Nigeria's domestic wheat has been from six to eight times what imported US wheat would have cost.

The above instances encompass the sub-Saharan African countries where irrigation development has been accorded high official priority, i.e., *privileged status.*

Other countries with a similar experience include Ethiopia, Somalia, Mauritania, Niger, Chad, and Tanzania. Except for a few commercial sugar estates (which have a better record), their achievements would not differ greatly from those we have reviewed. The main exceptions come from the areas of Africa where large-scale commercial producers have evolved their own irrigation systems, usually to protect export crops like coffee, tea, or tobacco. In Zimbabwe, Swaziland, and Kenya, there are producers who inherited or purchased highly efficient and cost-effective irrigation systems, a demand-led outgrowth of large-scale commercial farming in a hazardous environment. The contrast between demand-led and bureaucratically-imposed irrigation performance is striking.

Why Privileged?

In practice, the access that irrigation has to privileged status releases those responsible for designing and implementing projects from the necessity of adapting their solution to local circumstances or from learning how to overcome obstacles. Entrusting African irrigation to technologically-oriented irrigation agencies has created a situation (not found in the private sector) where each of the main parties--the government, external consultants, the irrigation parastatal, and farmers--can blame others for the systematic malfunctioning so widely evidenced, without doing anything much about it.

Our review of the African experience has described a wide gap in most instances between government objectives and the actual performance of modern irrigation. One explanation would be that this shortfall is the natural consequence of the first-generation *teething* problems which any new technology must undergo. To the contrary, we have argued that irrigation technologies represent bad investments for African development precisely because of their intensive, bureaucratically controlled, quasi-industrial character. When implemented in marginal environments within economic systems lacking imported inputs or where specialized production incurs higher risks, modern irrigation is *always* problematic.

This is a controversial conclusion, so let us examine in greater detail the reasons why irrigation remains attractive, particularly to governments, despite its poor performance on individual projects.

To combat drought is indoubtedly a major cause of African willingness to fund officially sponsored, modern irrigation projects. The vulnerablility of indigenous, small-scale irrigation in dry years is obvious. It is easy to assume that large-scale, modern irrigation is not subject to the same constraints, and thereby safeguards food security in dry years. For example, when after the Sahelian drought of the early 1970's, the six most afflicted countries banded together to form a permanent investment planning organization to combat drought (CILSS), they identified some 97 first-generation irrigation projects for implementation. Donors were reluctant to finance such a massive program totalling almost a billion dollars and designed to raise modern irrigation in the Sahel from 80,000 ha (in 1976) to 600,000 ha (by 2,000) (Somerville, 1986).

Donor caution was well justified in the Sahel. In the 1983-84 drought, modern irrigation sometimes failed those who had come to depend upon it. Furthermore, the large-scale, upstream developments located around attractive dam-sites shift the locus of water consumption away from the driest zone where it is most needed. The opportunity cost of diverting water into the inefficient *Office du Niger* is, for example, to further reduce the diminished downstream supply which serves farmers in more remote areas. The timing of irrigated production competes directly with farmers' own food crops, so that in good years, rice grown under irrigation occurs at the expense of diminished millet and sorghum output from off-scheme lands. Outside the Sahel, there is the additional objection that the crops irrigated (such as cotton and sugar) are for export, not food: if a real drought occurs, project farmers may be almost as disadvantaged as those without modern irrigation.

Saving foreign exchange spent on food imports and increasing food self-sufficiency are thus, chimerical goals. As the northern Nigerian schemes illustrate, production of wheat or maize under modern irrigation can become extremely expensive. Foreign exchange is lost at every step in the process: to pay for the initial design and construction, to obtain expatriate management (where required by donors on larger schemes), to purchase inputs such as insecticides and fertilizers, to maintain equipment, and to transport the crop once it has been harvested. At any more remote site in Africa, the hard currency component in irrigated production of hard grains other than rice usually exceeds landed import costs.

The notion that expenditure can be *saved* by investing in modern irrigation is perverse in other respects, too. It overlooks the generally high per-hectare investment costs in Africa. Higher African irrigation costs are partly caused by the components in the construction process. FAO figures indicate, for example, that in countries like Burkina Faso or Mauritania, the basic ingredients like cement, steel, diesel fuel, and engineering expertise, cost from two to three times or more what they would cost in India. FAO's estimated construction costs for a *full control*, lowland rice scheme such as one finds in Mali are, in 1985 dollars, about US $8,400 per hectare (FAO, 1986c). However, this same source notes a World Bank review of irrigation in Ghana which found some projects costing as much as US $40,000-US $50,000 per hectare. Nigeria's Bakalori scheme is estimated by those involved to have cost N 21,500 per hectare; in general, a Nigerian surface system based on a dam, costs between N 18,000 and N 22,000 per hectare (Maurya & Sachan, 1984).

Toksoz (1981) gives the cost of Kenya's Bura West project as over US $25,000 per hectare in 1980 prices. This may seem high when contrasted with the World Bank's large assisted projects in the Sudan, but is, in fact, less than half the cost of Kenya's own Turkana schemes as indicated in unpublished FAO calculations based on Kenya Ministry of Agriculture data (about US $63,000 per hectare). For further discussion of the northern Kenya experience with irrigation, see Richard Hogg (1983).

When costs are so high, irrigation agencies and donors have cause to disguise, in effect, the true size of their commitment. Officially reported figures are often incomplete, frequently overlooking preliminary survey costs, infrastructural investments, management charges, the decreased area actually put under irrigation, reduced cropping intensities and yields, and various unattributed cost overruns. Farmer labor costs have been assumed to be zero, even though this is their largest production input, and because of strong seasonality, there is a pronounced labor bottleneck during the cropping season. A Wageningen team study of eleven EDF-financed African irrigation projects found that most aimed at a 200 percent cropping intensity (two crops each year), which in practice was rarely achieved. In all cases, appraisal documents had shown an economic internal rate of return (EIRR) of 10 percent or higher. Project planners incorporated "excessively optimistic estimates of development possibilities, of production figures, and of implementation costs" (van Steekelenburg & Zijlstra, 1985:65-6). When recalculated on a more realistic basis, the EIRR became negative on four of the eleven projects, and with one exception (where only a small weir was funded for a pre-existing community system), the highest three did not reach a 10 percent level. The Wageningen team noted (van Steekelenburg & Zijlstra, 1985:28-9):

> In none of the projects that we evaluated was anything like such an assessment made before funds were committed. The general story is that of a . . . plain with a river running through it, which according to engineering standards could and should be converted into the *granary* of the country. And so work began!

Because of high investment costs, irrigation agencies feel impelled to impose single channel marketing to facilitate cost recovery, but this, in turn, entails growing crops like cotton or rice which are less attractive to farmers. Calculations by Simmons (1984) show that the gross margins farmers in Niger receive per day from irrigated cotton-growing (805 CFAF) are less than one half those obtained from rain-fed cowpeas intercropped with millet (1,784 CFAF), and a quarter those from irrigated, small-scale onion growing (3,290 CFAF). Rice returns are still lower. A Nigerian study finds that under large-scale improved husbandry, cotton gave the next-to-lowest return of twelve crops investigated (N 313 per ha); wheat was only slightly better (N 393), in contrast to onions (N 4,308), tomatoes (N 2,890), sugarcane (N 2,668), cassava (N 2,402), or yams (N 2,040) (PRC [Nigeria] Ltd., 1982). These figures support the deduction that African governments have chosen the crops to be irrigated on the basis of macro-economic, national interests, not farm-level returns (Faki, 1982). Palmer-Jones (1984) gives an insightful analysis of how tenants' and managements' pre-conceptions reinforce low performance. As Pearson (1980:21) observed on Sudan's New Halfa Scheme, "It is not uncommon for the nomad tenants to let their animals on to their cotton fields after only one picking ... as they feel the extra returns to be gained from the next pickings do not justify the labor costs."

Nevertheless, because irrigated farming seems modern and technologically sophisticated, and because most multi-national (but not bilateral) agencies have been willing to fund such projects, it retains its appeal to many African policy-makers. Their faith in the technology blinds them to its contradictory aspects in African settings. They have failed to see that, because of its high costs, it cannot be used to grow food crops, while at the same time, the bureaucratic controls required for cost recovery make it still more expensive. It is, in short, a *solution* designed to fail (Palmer-Jones, 1981).

Policy Implications

For the reasons we have reviewed, irrigation is no longer a privileged technology among most bilateral donors (Carruthers, 1983). Several African governments have unfunded major projects; for example, Tanzania's hopes to irrigate the Rufiji delta or Somalia's Juba valley. A consensus has emerged that donors should focus on rehabilitating older systems before embarking on new ventures; that small-scale, community-managed systems should take precedence over large-scale, parastatal ones; and that negative health impacts require close monitoring. Most also accept that African institutions for managing irrigation require strengthening.

However, the manner in which these second-generation projects are being carried out suggests that they represent a tactical retreat rather than a change in battle plan. Small projects can be just as bureaucratic and inefficient as large ones as, for example, indicated in Mali's Action Bl-Dir Project (Moris, 1984). Donors have funded rehabilitation projects which focus on physical reconstruction, while ignoring the organizational changes to ensure that maintenance will be carried out in the future. Some large donors have shown themselves astonishingly inept when trying to operate on a small-scale, as, for example, in assisting West African swamp rice projects (Dey, 1984). It is clear that merely switching from new projects to rehabilitation and from large to small is not enough (Underhill, 1984; Blackie, 1984; Wensley & Walter, 1985). African irrigation projects must undergo a basic transformation in modes of approach: from a *blueprint* to a *process learning* strategy, and from bureaucratic to participatory management. Neither aspect is part of the received technology for project implementation as it is currently employed within Africa.

The loss of privileged status opens the door for irrigation planners and engineers to learn from the African experience. It would appear from the larger study on which this article has been based that African environments diverge from the orderly setting presumed in official plans in at least five respects.

First, the high evaporation potential and rapid run-off in African river systems make the security and adequacy of the water supply a critical consideration. African dry lands present few medium-sized supplies which are amenable to irrigation development, a situation which reinforces the marked dualism between large-and small-scale systems.

Second, the extreme seasonality of peasant production in dry environments results in the paradox that, while for much of the year people may be underemployed, in the cropping season the labor force is usually fully occupied. Any activity superimposed on rural households during this period occurs at a high opportunity cost in terms of food production foregone (Longhurst, 1986). Irrigation planners should expect strong competition between irrigation activities and other uses of farm labor--a situation which requires close attention to crop returns and labor demands when planning irrigation.

Third, farmers' own production strategies diverge sharply from what irrigation planners assume they should be. In many farming systems, household heads safeguard their incomes by extensive cultivation, diversified enterprises, low capitalization, low use of purchased inputs, low reliance on official services (because of their unreliability), and by retaining a high degree of flexibility over the course of each season. In contrast, irrigation schemes have expected farmers to specialize, to intensify production, to make heavy use of purchased inputs, to depend on government services, and to maintain improvements they do not own themselves. In a local context, this strategy decreases farmers' food security and increases their vulnerability to adverse external events.

Fourth, land tenure is problematic almost everywhere in sub-Saharan Africa. Planners have relied upon land laws inherited from colonial powers or sometimes military decrees to obtain possession of the area needed for irrigation schemes (Francis, 1984). The idea that such lands are vacant or unused is a convenient fiction. In actuality, implementation of any large project will usually require the dispossession and resulting impoverishment of many households, perhaps more than the project can subsequently accommodate (Tiffen, 1985). In any planning of future schemes, close attention should be given from the very start to existing resource utilization and to advance arrangements for compensation to those affected (Bloch, 1986).

A fifth and final difference relates to the managerial environment. Western donors these days are neither sympathetic to, nor cognizant of, the particular managerial problems faced by African parastatals. Such agencies operate within a quite distinctive matrix of possibilites, not handled under classical management theory. Tribalism and corruption are perhaps widespread; there are tremendous pressures to add unnecessary staff; and the agency must compete with others by offering superior fringe benefits (since salary levels are usually fixed nationally).

A similar gap occurs in water management training in overseas countries. Western experts customarily equate irrigation management with hydraulic management at the perimeter level, and on-farm water management at the field level. While both are significant in any irrigation system, they are only part of what concerns an African irrigation manager. Sustaining higher output over the longer run depends upon management's ability to support an introduced technology for which water is only one of several inputs. There are many aspects of African irrigation management, (ensuring input supplies, maintaining equipment, obtaining cost recovery, controlling disease outbreaks, scheduling activities, etc.), about which conventional irrigation management is silent. At issue is not what the water engineers recommend, but rather the critical topics they ignore.

Lessons for Participation

What specific observations does the experience we have reviewed suggest in regard to the prospects for increased farmer participation in Africa's third generation projects? Let us summarize the main lessons presented in detail elsewhere (Moris & Thom, 1990).

(1) *Determine where irrigation is inappropriate.* It is obvious in retrospect that irrigation should not be located where farmers are primarily subsistence-oriented, or in areas remote from centers of demand. The technology is heavily dependent upon reliable support: equipment suppliers, spare parts, mechanics, good seed, fuel, transport, fertilizers and insecticides. In many poorer communities, these key components are only occasionally available to African farmers. Small irrigation projects rarely have the resources to duplicate or compensate for unreliable production services of a more general nature. And yet, these are essential before high volume, intensified production can be made profitable. When projects are implemented without considering these institutional risks, farmers who participate will find themselves incurring large debts without receiving more secure sources of food or increased profits. They are then likely to resist adopting any technology which increases their dependency upon erratic and unreliable external institutions. Irrigation planners are accustomed to evaluating the riskiness of rainfall; they need to learn how also to judge institutional riskiness. In Africa, problematic upstream and *downstream* linkages are a key reason for the low performance of many smaller schemes. Participation in what farmers regard as loss-making ventures will always be difficult to promote.

(2) *Smallness alone is not enough.* As indicated above, the switch to smaller projects does not necessarily lead to greater participation by farmers. Here Wensley and Walter (1985) have pointed out that the problem lies not so much with the techniques as such, but rather in employing the same systems architecture used on large projects. Most donors continue to rely upon an elaborate project cycle when planning new schemes, and upon formal management once they are in operation. As usually carried through, the *project cycle* leaves initial data-gathering to design experts. *Participation* means to them that once the project is fully outlined (and even after field layouts are determined), local farmers are given a chance to review the plans and make minor amendments. This is essen-

tially token participation undertaken as a sop to public opinion so that planners can claim that they have the full backing of the farmers who will, after all, carry out the actual production activities. Such projects are often just as bureaucratic and rigid as large ones, and--as we have pointed out--they may have even higher unit development and operating costs.

The differences in *systems architecture* between pre-planned and evolutionary approaches to small projects can be represented in schematic form, as indicated in Table 5.1 (Moris and Thom 1990).

(3) *Whether to have salaried staff.* Some salaried staff will be required in almost all forms of irrigation development since even small projects can require expert assistance on occasion. Here, however, we refer to on-site staff who are employed to manage water delivery and (sometimes) the production activities as well. As a rule of thumb for African irrigation, one can state that the more employed staff a scheme requires, the less the chances for genuine farmer participation. The reason is partly a matter of bureaucratic hierarchy: employees will naturally look to their paymasters for instructions. Since smallholders have an extremely limited capacity to pay staff salaries, any project with a sizable staff nucleus will take its orders from above and not from below. It is that simple.

The problem of scheme staff is heightened by the fact that in Africa, going back to colonial days, one often finds a highly differentiated structure for low level occupations. A group of farmers cannot simply hire one person who will provide the various skills which pump operation and water distribution require. If they could, it would be feasible to talk about user groups who employ their own support staff. Instead, one finds pipe fitters who know nothing about pumps, pump mechanics who are not qualified to drive a truck, truck drivers who are illiterate and cannot read an order sheet, and water guards who cannot drive. Even a quite simple project will often require 10-15 low level salaried staff, creating an intolerable burden upon farmers if they are expected to meet each month's salary claims.

(4) *The cost recovery problem.* On actual site visits, one is told again and again that the reason why African irrigation schemes have stressed the controlled marketing of export crops like cotton has been to facilitate cost recovery. If an irrigation scheme employs a cluster of staff and operates a mechanized pumping system, it will have substantial recurrent costs. Where farmers are left to grow crops of their own choice, they have been unable or unwilling to repay scheme charges. To salaried scheme managers (and to ministries of agriculture), the answer has seemed

TABLE 5.1 Conventional Versus Evolutionary Approaches to Irrigation Development

Conventional Pre-planned Development	Evolutionary Small-scale Development
Top-down approach	Bottom-up approach
May derive from a larger Master Plan	Evolves out of perceived local needs
Little pre-design or pre-construction local involvement	Often intensive discussions and debates preceding commitments to project
Project selection by government and external donor	Activities not seen as constituting a development project
All major decisions made by government agencies	Major decision made by various interested parties
Data intensive design procedures	Minimal formal acquisition of planning data
Design done at distance by specialists	Design done in locality, by contractors
Blueprint plan arrived at to guide all implementation activities	No overall plan to guide implementation activities
Scheduled within fixed time frame	Open-ended time frame for gradual development
Large initial investment physical works	Major construction often in mid-way in system
Low initial utilization, rapid deterioration	High initial use of fairly modest physical works
Users expected to conform to planned layout	Users improve and adapt the systems as they go
Facilities completed at outset and secure if properly used	Facilities subject to catastrophy such as loss of headworks or main canal
Physical reorganization of fields and farms often required	Minimal reorganization of fields and farms in gradual development of system

Source: **Moris and Thom (1990:389)**

to be the imposition of single channel marketing of some crop which the scheme can tax before paying farmers back for what they produce. The farmers, too, prefer to be taxed in this way, since then they never see the money they are contributing, and thus find it easier to refuse the claims of debtors, friends, and relatives upon their income.

From a participation standpoint, we can see that scheme managers tend to accumulate more and more power (Chambers & Moris, 1973; Barnett, 1977). When international commodity prices fall, as they have for sugar and cotton throughtout the early 1980's, the irrigation bureaucracy will be tempted to squeeze farm level profits before reducing the management's own costs. It is only a small step further to insist that farmers should be *tenants* rather than plot-owners, giving management a theoretical power to evict those farmers who fail to meet management standards. In many African irrigation projects, *participation* is defined from a managerial perspective as being a tenant's obligation to grow stipulated crops, to clean the canals, and to pay required charges. This is not, of course, how farmers see the matter--leading producers to evade formal controls while carrying on a range of economic activities which do not officially exist.

The general lesson is that irrigation planners need to find other methods for cost recovery. Where controlled marketing and scheme residence are imposed upon farmers, the resulting scheme organization is bound to become autocratic and over-centralized. It will see farmers' legitimate needs and requests as being a threat to management, and farmers, in turn, will fail to identify with the scheme's production objectives. In the longer run, the linking of water delivery to controlled settlement is unnecessary and often self-defeating. Settlement schemes have an array of additional problems requiring skilled assistance beyond what an irrigation agency can usually provide.

(5) *The desire for increased control.* In a turbulent environment where many things can go wrong, managers often feel a need for a high degree of control over actual production activities. In reality, of course, most irrigation project managers control very little once the project is in full operation: the hours for pump operation, and perhaps when and where water gates are opened and closed. In Africa, the reason ministries of agriculture often create intermediary scheme units for each perimeter has been to increase the apparent control exercised by the management over irrigated production. And yet, these same officials also claim they want greater farmer involvement. Most African irrigation schemes have

some type of farmers' committee to assist in water delivery scheduling and to convey farmers' views about scheme operations. However, few irrigation officials have been willing to recognize that having a high degree of managerial control over field operations is antithetical to having enthusiastic farmer participation. One can assume that farmers will participate in direct proportion to the degree of control they feel they exercise, not only in paying the costs of operation, but also in receiving immediate benefits.

The Search for Better Models

It seems that the reasons for low farmer participation in many African projects are interconnected, and are of a structural nature which transcend immediate on-site considerations. If irrigation projects continue to be funded by external donors acting through public agencies, if the mode of planning emphasizes the project cycle (which puts most investment and preparatory work at the start), and if salaried staff are appointed to supervise what farmers do and to extract repayment of the loan, we can predict that the low degree of performance seen in the past will persist into the future. Field staff will continue to show low commitment, and will suffer from poor morale. Farmers will find they have little meaningful say over what crops are grown and how the irrigation scheme is run. Of course, in individual cases, an unusually hard-working manager and sympathetic farmers may temporarily raise scheme productivity to the anticipated levels made possible by the improved technology. However, over the longer run, the kinds of problems described in this chapter will reappear, and overall performance of small-holder irrigation is likely to be disappointing.

This is, then, the fundamental reason why irrigation should not be treated as if it were a *privileged technology,* a proven answer to farmers' needs which can be copied without modification from approaches employed in the large farming context of American or Australian farming. To gain a greater level of commitment from African small-holders will require developing a different *systems architecture,* which takes into account the water supply problems seen in the Tropics and also the reduced leverage which small-holders can exercise over the irrigation bureaucracy.

6

Irrigation in the Western United States

Pamela J. Riley

Introduction

In this chapter, the historical development of irrigation systems in the United States is analyzed by looking at governmental involvement and at U.S. water law. The next section presents the three levels of American irrigation organization and examines their relationship to each other. It includes suggestions for promoting efficiency and incentives for participation. The final section is a discussion of the major contemporary problems facing irrigated agriculture in the nation and includes some potential solutions for dealing with these problems.

Historical Background

Irrigation in the United States may be traced back to the Hohokam Indians in the southwest who effectively built and used extensive canal systems; a canal dating back as early as 300 B.C. has been uncovered. They built adobe-lined canals totaling more than 135 miles and irrigated 100,000 acres of beans, grains, melons, squash and cotton (Dunbar, 1983; Campbell, 1986). The Hohokam civilization reached its peak about 1200 A.D. and then disappeared by 1400 A.D. Research suggests they were forced to abandon their fields because of salinity problems (Worster, 1985). Other Indian groups, the Pimas and Subaipuris, succeeded them. Later, the Spaniards, who had been irrigating since Roman times, established irrigation in New Mexico, Texas, and California (Dunbar, 1983).

Beginning in 1847, the Mormons were the first Americans of Northern European ancestry to practice large-scale irrigation. By 1902, they had six million acres under irrigation in Utah, Arizona, California, Wyoming, and Idaho - the same year the U.S. Bureau of Reclamation (BuRec) was established by the federal government (Reisner, 1986).

The Reclamation Act of 1902 empowered the Secretary of the Interior to "locate, construct, operate, and maintain works for the storage, diversion, and development of waters for the reclamation of arid and semi-arid lands in the Western States" (Powledge, 1982:11). Ostensibly the BuRec was to construct large-scale projects for small farmers. This Act permitted the federal government to develop water projects while water was state regulated or owned.

Through the extremely cost intensive efforts (exceeding 11 billion dollars [Worster, 1987]) of the BuRec, incredibly high and large dams were built on rivers in the driest portion of the country for purposes of irrigation and generation of electricity. By the 1960s, the BuRec, along with the U.S. Army Corp of Engineers, had constructed large dams in almost every possible site in the West. The result was tremendous economic development and population growth in California, Arizona, Colorado, and in otherwise inhospitable areas. For example, in the 70 years from 1910 to 1980, Los Angeles grew from 319,000 to 3,122,000; Phoenix grew from 11,000 to 772,400; and Denver more than doubled (Dunbar, 1983).

Although irrigated acreage presently consumes only about one-seventh of the nation's cropland, it produces more than one-fourth of the value of American crops because average yields are so much greater on irrigated land. Approximately 83 percent of the irrigated acreage in the U.S. is located in the 17 western states (Frederick & Hanson, 1982). According to the last agricultural census, the West contained over 45 million irrigated acres (Worster, 1987). Agriculture accounts for about 88 percent of water consumption in the West (Frederick & Hanson, 1982).

Surface Water Law in the U.S.

Initially, American water law was adopted from English Common Law and based on the *Riparian Rights Doctrine*. Under this doctrine, anyone owning land along a stream or river has an equal right to divert that water for personal use but must flow it back when finished. Thus, the right to water is restricted to those whose land borders a watercourse. Under this system, no one is allowed to *kill* the stream by diverting or impounding its *total* flow.

Such a system works well in areas with a great deal of precipitation and large numbers of tributaries; however, it was ill-adapted to the conditions found in the arid areas of the American west. The Riparian Rights Doctrine continues to operate in the east, but by 1850, the western states began to reject or modify it.

Currently, in the western U.S., owners of riparian property generally do not have a right to the water flowing through their land. The *Doctrine of Prior Appropriation*, or the *first in time, first in right* principle was developed to stem conflicts between rival irrigators and to provide them with greater security. Thus, as with a mining claim, the first person laying claim to the water for a particular use has priority over other claimants.

Senior claimants may divert the water they need, even if it means those who come later receive no water--provided that water is put to *beneficial use.* Consequently, early in the spring when there is usually plenty of water, the rights of all users are satisfied--but later in the season as the stream flow decreases senior appropriation rights are served first and junior appropriators may receive no water at all. Appropriative rights may be forfeited if a user discontinues use for a certain period of time, determined by the state--usually three, five, or ten years.

The *Beneficial Use Doctrine* refers to what is deemed as a legitimate use of the water. This would include such uses as: irrigating a crop, watering livestock, maintaining a fish pond, putting it to industrial or domestic use, and watering a lawn. In many western states instream flows to protect fisheries and wildlife habitats are now recognized as benefical use. Waste is not considered a use and a person cannot take more than they can put to beneficial use. If beneficial use is discontinued, the right is discontinued. One cannot lose the right to use water without a court decree.

Water rights in the West are property rights which can be bought and sold, but one only owns the *right*--not the water. One does not buy the water but pays for the cost of management and the right to use it.

Three Levels of American Irrigation Organization

Approximately 75 percent of surface water is supplied through some kind of water organization. Organizations were developed to facilitate delivery of water from water sources away from the farm. Unincorporated mutuals, incorporated mutuals, and irrigation districts account for about 95 percent of all irrigation organizations in the nation (Frederick & Hanson, 1982).

To understand irrigation in the U.S. it is useful to examine the three levels of organization and how they interface with and relate to one another. Each level operates with emergent joint agreements and has its own rules (the social aspect) and tools (the technical characteristics). Uphoff (1986a) suggests that irrigation management includes: (1) control structure activities: design, construction, operation and maintenance of systems; (2) water use activities: acquisition, allocation, distribution, and drainage of water; and (3) organizational activities: decision-making and planning, resource mobilization and management, communication and coordination, and conflict resolution. These are all significant considerations in the organizational tasks associated with water management and are relevant to all levels of irrigation organization.

The Main System

The first level is the main system, the central organizational bureaucracy which manages large-scale works and is responsible for water capture and delivery. At this level, large volumes of water must be conveyed within boundaries. The main system focuses on the watersheds, construction of dams, catchment areas, storage or diversion facilities, and canals. They determine "watershed yields, storage capacities, interactions of ground and surface water flows, and physical structures for control, measurement, and drainage" (Freeman, 1986:11).

The main system incorporates both federal and state bureaucracies. At the *federal* level, the primary agency is the BuRec which delivers approximately one-third of the surface irrigation water used in the West. It is an arid-land agency only operating in the 17 western states with commissioners in Washington, D.C. and 7 regional offices. Each regional office has several project offices with an engineering division for water studies, a water and power control division for water scheduling, an administrative services division, and a water and land operations division. The BuRec deals with three organizational types: water user associations, irrigation districts, and water conservancy districts.

Also at the federal level, the U.S. Department of Agriculture (USDA) works at the farm level through several agencies such as the Soil Conservation Service (SCS). The SCS coordinates its efforts with the BuRec and provides free assistance to land users through nearly 3,000 locally run soil conservation districts. These are local units of government generally organized along county lines, and led by a governing board of local farmers, ranchers, and community leaders. The SCS emphasizes soil and water conservation, natural resource surveys, and community resource protection and development. It provides technical assistance by giving on-farm consulting and farm layout and crop assistance. The farmer discusses individual soil and water management needs and then makes her or his own decisions based on the alternatives presented. In some instances, the SCS provides financial aid, often in the form of cost-sharing. The Agricultural Stabilization and Conservation Service (ASCS) is also an agency of the USDA and provides funds for farm improvements (e.g., lining ditches) approved by the SCS.

At the state level are Water Conservancy Districts, and river basin and valley projects (e.g., the Central Arizona Project). Water Conservancy Districts are formed to put water to beneficial use. In the past, it was primarily agricultural use but municipal uses are becoming more common. They are generally quasi-municipal corporations and belong to the people who live within the district; it is paid for through property taxes, water assessments and water bills. The Northern Colorado Conservancy District provides an example, it may:

> acquire water; obtain rights-of-way for works; contract with the United States or otherwise provide for construction of facilities; assume contractual or bonded indebtedness; administer, operate, and maintain physical works; have authority to conserve, control, allocate, and distribute water supplies for supplemental use; and have taxing and contracting authority to derive the revenues needed to accomplish its purposes (Barkley, 1981:3).

Thus, the people in a region of the state have an agency to work on behalf of their particular interests. The courts or the governor of the state appoints the Board of Directors to ensure equitable representation.

Because of their mandate to construct and implement large-scale projects, main system level organization must address *average* needs; rewards are not associated with the productivity of a specific farm. Management requires an understanding of general principles and the primary objective is overall operational smoothness (Freeman & Lowdermilk, 1985).

Local Command Area Level

The middle level is the local unit command area organization. In a large-scale system, water control is shared by government bureaucracy and the farmers; because these two groups have different objectives, it is necessary to have a buffer between them. The unit command system provides this interface between the farmer and the main system and focuses on disaggregating water from the large volume main systems into appropriate amounts for on-farm application. The domain of the unit command area is the distributaries and watercourse which deliver the water to the farm gate for field application.

The emphasis is on facilitating farmer control of water appropriate to unique on-farm needs. Middle level managers answer both up and down the system, attempting to reconcile the agendas of both the main system and the farmer. Therefore, middle level organizations must be carefully designed to deal with issues of concern, both to farmers and the upper level bureaucracy. A proper interface allows both the main system and the farmer to pursue their own concerns.

Middle level organizations take the form of Water User Associations (also called mutual, irrigation, or water companies) and include anywhere from six to eight farmers to hundreds. They may be river-, groundwater-, or reservoir-dependent, or some combination thereof.

The simplest organization is the unincorporated mutual company which typically provides water to seven or eight farms whose farmers build a simple diversion on a mutually held stream or who commonly build and maintain a canal. Agreements may be informal and verbal. These companies constitute over half of the irrigation organizations but supply water to merely 11 percent of the land irrigated by all organizations (Frederick & Hanson, 1982).

Incorporated mutuals are larger and more formal. They account for about one-third of all the organizations and serve slightly more than one-third of the irrigated land (Frederick & Hanson, 1982). They are usually nonprofit organizations which issue stock to the farmers under the system. Stock is not equivalent to the land owned.

Shares may differ when size of holdings do not; therefore, some may have enough water rights while others are inadequate. One might sell rights to another and, thus, redistribute water in the market place. Examples of types of shares include: (1) receipt of a percentage of the volume of water available from the company; (2) time period rotation where an individual is allocated a particular time each week; (3) a turn procedure whereby the first farm on the canal takes all the water needed and then turns it over to the next farm, and so on; (4) a farm priority system in which the earliest settled farms are the first served, each farm taking what it needs before turning to the one next in priority; (5) a crop priority system based on the economic value of the crop; (6) a demand system allocates a fixed quantity of water to each farm for the season and the farmer determines how much to utilize each irrigation application; and (7) a market system where users bid for the water they desire for each irrigation period (Maass & Anderson, 1986).

Incorporated mutual companies dig ditches, hire people to monitor ditches (ditch riders), and deliver water. Generally, the state *owns* the water and individuals pay the cost of recapturing and delivering. Such companies are generally composed of a Board of Directors, various subdirectors, a manager, and maintenance personnel--all responsible to the water users who are shareholders in the system. They may also have a contractor and engineer for land and/or water purchase and an attorney.

Irrigation districts account for 10 percent of the organizations but supply water to nearly half of the land irrigated by organizations (Frederick & Hanson, 1982). They are self-governed and financed public corporations organized by law. They have the power to tax nonwater users as well as users. All landowners in the organization have one vote, regardless of how much land they own, and the government designates the land which is accorded water rights.

Clearly, effective institutions at this level are extremely important. Farmers will support associations if there are incentives to do so and if they perceive the system as equitable. For example, head/tail problems must be considered in the design of the share system. Farmers are unhappy when they have to absorb the costs of poor control over water which results in waterlogging, salinity, and so forth. The belief that an adequate and reliable supply of water will result from farmer participation in the operation, maintenance, and rehabilitation of irrigation schemes is a major influence on the decision to participate (Uphoff, 1986a). Research suggests that farmer involvement in planning, design, water allocation, and conflict management tends to increase commitment. However, a lack of direct participation in WUA's may occur precisely because they are working and people are satisfied (Lusk & Parlin, 1988). This is not at all uncommon in the American West.

Conflict in organizations can occur around land disputes, watercourse location, water theft, bribing of officials, and illegal purchasing of water.

Farm Level

The third level of irrigation organization is the farm. The local command unit passes the water on to the farmer who controls the water on-farm. Here, good management is measured by the ability to apply high quality water to crops at the right time and in adequate amounts. Ultimately, however, control at this level is dependent on the organizational networks at the other levels. As Freeman (1989) suggests, it is the organizations developed to rehabilitate, operate and maintain irrigation systems and to settle conflict at all three levels that determine the effectiveness with which the crop root zone requirement is met.

At the farm level, unique conditions are a major consideration. Field productivity is the primary concern and the objective is to adapt as quickly as possible to changing conditions--from field to field, crop to crop, and season to season. Success is not only contingent on the efficiency of middle and upper level organizations, but also requires knowledge of local conditions.

Special Problems in American Irrigation

Government Involvement

The activities of the BuRec and Army Corp of Engineers have come under considerable scrutiny in recent years. Reisner (1986) points out that in their competition with one another and in their zeal to build as many dams as possible, some of the dams built actually drown more land than they irrigate and many do not produce enough crops to ever pay for themselves. Some dams have also provided water illegally to agribusinesses too large to qualify for the Bureau's subsidized water. All of this occurred despite the fact that the U.S. Congress was well aware of reports, on a number of occasions, which questioned the economic and/or environmental soundness of construction. These are cases of pork-barrel politics at its worst. Water projects have consistently been used in the U.S. Congress to trade favors and, with the pressures from their constituencies, few congressional representatives have been willing to oppose *any* proposed water projects.

Recipients of BuRec water are supposed to reimburse the government through fees charged irrigation districts for the cost of the water, but this has rarely been the case. The initial period of payment of interest-free indebtedness was set at ten years, later changed to twenty years and later yet to forty. Stegner (1987:18) notes that:

> Eventually, much of the burden of repayment was shifted from the sale of water to the sale of hydro-power, and a lot of the burden eliminated entirely by the practice of river-basin accounting, with write-offs for flood control, job creation, and other public goods.

Ultimately, BuRec programs have not supported the small family farm but have subsidized the cost of irrigation water for large agribusinesses. To receive subsidized water, the original legislation required a farm to be no more than 160 acres for each family member and the family had to live within 50 miles of the land. The BuRec has often ignored violations (some farms include thousands of irrigated acres) and has never enforced this law.

The conflict today is over the fact that taxpayers throughout the U.S. are required to subsidize western agricultural reclamation. People in many regions of the country resent what they perceive as local farmers being forced into unfair competition with farmers in the West who produce subsidized crops which are often in surplus elsewhere. In other words, people have been subsidized to grow crops that people in other regions were being paid *not* to grow. Because of these conflicts, newly proposed federal water projects are no longer getting much support from Congress.

Water subsidies. Market forces have had very little to do with decisions about which water projects to pursue. Water has been a highly subsidized commodity in the West, resulting in profligate and inefficient use. Water for agriculture has been subsidized, not only by the federal government, but also by power revenues (sometimes absorbing 90 percent of the cost [Reisner, 1986]), municipal revenues and property taxes. Despite the fact that water is becoming more scarce and valuable, it continues to be treated as a free good--the cost amounting simply to its capture and transport from its source.

Some argue that bringing water into the marketplace would be a good method to eliminate the water shortages in the West. Legal rights to water could be sold and traded by the acre foot. Welsh (1985) reports that California and Utah have proposed a water bank by which potential buyers could make bids which farmers consider in determining whether to utilize or sell all or part of their water that year. Possible results include a greater inclination to retire land not suited to agriculture, more efficient irrigation systems, more low water consumptive crops, and less pressure for the building of expensive and environmentally destructive dams.

Under existing western water law, the *first in time, first in right* system does not allow, in most states, the holder to sell the excess water. As mentioned previously, if one does not use a water right within a certain period, the right is lost and may be appropriated by another. The reality is one of *use it or lose it.* In short, there are no incentives to conserve water under the present system; thus, there is little hesitation to produce crops that are highly water consumptive. Given that much of the value of desert land is tied to access to water, people have an incentive to grow water consumptive crops in an effort to capture the highest water right possible.

Nonetheless, the free-market transfer of water is not necessarily a perfect panacea to the problem. Water marketing is likely to make the large agribusinesses, which obtain their water via dams funded by the taxpayers, beneficiaries of large windfall profits (Udall, 1987). Many small operators would be unable to compete for water on the open market against corporate agriculturalists and city industrialists.

Conflicts Over Water in the West

Sources of conflict among users are many. As Frederick and Hanson (1982:2-3) have observed:

> Western water use and development have become issues of growing national concern and debate--often pitting one region of the country against another, or western farmers, conservationists, and developers against each other. Development of western energy resources, expansion of western urban and rural areas, preservation of the west's natural resources, continuation of federal subsidies for western water development, Indian rights, as well as the role of western irrigation in U.S. agriculture, are at issue in the resolution of these conflicts.

Irrigation in the southwest has created vegetation in excess of 3.4 million acres of desert which produces crops worth over $1 billion (Boslough, 1981). Farmers and ranchers currently use almost 90 percent of the only significant surface source of water in the southwest--the Colorado River. This river is "the most legislated, most debated, and most litigated river in the entire world. It also has more people, more industry, and more significant economy dependent on it than any comparable river in the world" (Reisner, 1986: 125).

The Colorado's headwaters are located high in the Rocky Mountains and lead 1,400 miles south to Mexico. By the time it arrives in Mexico, it is not much more than a brackish creek--undrinkable and too saline for agricultural use. The Colorado River Compact of 1922 guaranteed 7.5 million acre-feet of water annually to the upper basin states of Colorado, Wyoming, Utah, and New Mexico; 8.5 to the lower basin states of California, Nevada, and Arizona; and an additional 1.5 million acre feet was designated to Mexico. The problem is that the decision on the amount of water to allocate was based on an unusually wet period in the history of the river. Consequently, 17.5 million acre- feet have been promised annually, but the actual average flow since 1930 has been about 12 million acre-feet (Stegner, 1987).

Central Arizona Project. The Central Arizona Project (CAP) is a $2 billion water delivery project that will receive 1.5 million acre-feet of Colorado River water which it will carry 300 miles across the desert to central and southern Arizona (Campbell, 1986). CAP water began to flow in 1985 and the project is expected to be completed by 1998.

Water conservation was one of the goals of the project because Arizona has a serious problem with groundwater depletion. The state requires that groundwater usage must be reduced by one acre-foot for each acre-foot of CAP water a farmer receives, and prohibits the irrigation of any new farmland with CAP water. The problem is that groundwater will be so much cheaper than CAP water that there will not be much incentive to purchase it (Reisner, 1986).

The major conflict is with California because the CAP will ultimately result in the demand on the Colorado River exceeding supply. California contributes no water to the Colorado, but takes the most; it has been drawing one million acre-feet more than its allotment for years. This happened under the Appropriative Rights Doctrine; California simply began to use Arizona's entitlement under the Colorado River Compact before Arizona did. With the completion of the CAP, Arizona will absorb its full entitlement.

In a lawsuit Arizona brought against California over Arizona's Colorado River Compact entitlement, California lost and had to forfeit the share of Arizona's water it had appropriated. All is not rosy for Arizona, however. Under a clause in the Compact, water shortages are not shared among the states. That is, California gets its full 4.4 million acre entitlement before Arizona gets any water; Arizona gets whatever is left. Because more water was allocated under the Compact than exists in an average year, in a year of drought Arizona may get *very* little water. Ultimately, then, the CAP may not be worth much to Arizona.

Native American water claims. When the states and federal government were allocating water resources in the West, they gave little thought to Indian water rights. However, the reservations constituted over one-fourth of the land in Arizona and the U.S. Supreme Court ruled in one case that the reservations must be guaranteed an adequate water supply to all irrigable land within their boundaries.

Rights predating the Colorado River Compact must be settled first. Claims by Indians have been made on the basis of seniority going back to the creation of the reservations in the late 19th and early 20th century. A large number of cases are presently in litigation and more can be expected. As these claims are met, the availability of water could change in many parts of the West--including the water supply of Los Angeles.

Conflict between municipal and agricultural uses. The rapid population growth in many western cities has resulted in increased demand for municipal water. Per capita municipal use of water is higher in the West than in any other area in the country. In the summer water demand is particularly high as people irrigate lawns, golf courses and parks in the desert, operate evaporative coolers, and maintain fountains and artificial lakes. An example is Denver, Colorado where 51 percent of the municipal water supply is used to irrigate lawns, accounting for 9 percent of the city's water consumption (El-Ashry & Gibbons, 1988). To maintain this lifestyle (and as long as water is subsidized, and therefore very cheap, there are not incentives to modify behavior), water availability is a major priority.

Many cities in the West have been purchasing farmland as a security buffer to allow them to transfer agricultural water to the metropolitan areas in the future. This will ultimately eradicate many rural areas that have been dependent on agriculture.

Recently some cities have suggested compensating farmers to line their ditches and to install more efficient irrigation systems to make the salvaged water available for municipal use. The problem is that most existing water law requires that water not put to *beneficial use* is not part of the farmers appropriated right and hence is available to other appropriaters. That is, it is not available for sale or transfer since it does not belong to the farmer. Reisner (1990) argues this is probably the greatest obstacle to encouraging water conservation.

Certainly, the rapid depletion of groundwater puts present users into potential conflict with future generations. In twelve western states, municipal and domestic water use is given the highest priority. In nine states, agricultural use is given second ranking (Frederick & Hanson, 1982).

Groundwater Pumping

Groundwater wells have burgeoned with the spread of oil-powered pumps resulting in rapid depletion of groundwater. This is a major problem in the nation. Hydrologists refer to it as *water mining* because the water is being pumped out of the ground far more rapidly than nature can replenish it. The million- year-old Ogallala aquifer is the largest in the U.S. and is located under some 20 million acres in Texas, New Mexico, Colorado, Oklahoma, Kansas, and Nebraska. Almost half of the nation's cattle industry, a quarter of its cotton crop, and much of its corn, grain, sorghum, and wheat is supported by the aquifer (Boslough, 1981).

Hydrologists predict the Ogallala is being lowered so rapidly that high plains irrigation will be over by the end of the century (Boslough, 1981) and that by 2020, five million acres of irrigated land will be dry (Little, 1987). In the western plains of Texas, the Ogallala is being depleted 18 times faster than it is being re-plenished (Boslough, 1981). In some areas, the Ogallala's *over-draft* has reached 95 percent with the only source of restoration being local rain and, occasionally, rain-swollen rivers. In Colo-rado and Kansas, water tables are dropping two to three feet a year and as much as ten feet in some dry years (Little, 1987). When the cost of energy skyrocketed in the late 70s making pumping prohibitive, tens of thousands of acreage in Texas, Colorado, Kansas, and New Mexico was returned to dryland (Reisner, 1986).

Arizona receives an average of less than 10 inches of rain a year and its annual average run-off is the lowest in the U.S., 0.4 inches (Folk-Williams, Fry, & Hilgendorf, 1985). Sixty percent of the water supply in the state is mined from the ground. Tucson, Arizona, a desert city with a population of over one-half million, is totally dependent on nonrenewable groundwater. They pump water from depths of up to 450 feet and in some areas the pumping depth is falling as much as 5 feet a year (Welsh, 1985). The water table in the Tucson area has declined by as much as 700 feet since they began taking measurements (Turner, 1987). Groundwater over-drafting is forcing Arizona into an urban economy.

Extensive groundwater over-drafting has resulted in major fissures and crevasses in the earth's surface. Ground subsidence can crack highways, dams, buildings, and sewer systems.

Environmental Concerns

Salinity. Much of the soil in the West is classified as saline or alkaline and the primary water quality problem is salinity. As irrigation water is reused, it continues to pick up salt and fresh water evaporates. With poor (i.e., slow) drainage, the salt builds up in the root zones, starving plants of oxygen and killing the crop.

The soil becomes waterlogged and damaged structurally--making it practically impermeable to water. Ultimately, this results in desertification and irreversible damage.

The Colorado River has a greater problem with salt than any other river in the nation. About half of this salt may result from natural sources but much of it comes from irrigation waste water that percolates into the groundwater system and then enters rivers.

Erosion/silting. In dry areas, irrigation can serve to protect soil from the wind by providing a greater vegetative cover. Land leveling also reduces erosion and may be done with the installation of an irrigation system. Further, in streams with heavy natural sediment loads, irrigation can reduce the quantity of sediment.

Unfortunately, there are also negative aspects. Extensive erosion occurs with the use of flood and furrow irrigation systems on land which is inadequately leveled. Gravity irrigation is the only form of irrigation used on 66 percent of the irrigated acreage in the West. This form is particularly likely to lead to erosion. When irrigation is terminated, the soil becomes very susceptible to erosion. The advent of center-pivot irrigation has resulted in cultivating highly erodible sandy and hilly soil. The wheel tracks of the large mobile sprinkler systems may initiate the formation of large gullies.

In 1977, about 35 percent of the irrigated cropland and pasture in the West was considered susceptible to erosion; 21 percent of the cropland and 28 percent of the pasture was classified as highly erodible. However, there is twice as much highly erodible land in dryland row crops as in irrigated (Frederick & Hanson, 1982).

Irrigation may contribute to erosion and sedimentation and to the needless destruction of wildlife habitats (Heuvelmans, 1974). The erosion caused by building roads and logging, which is threatening some of the country's highest quality water and fisheries with sediment (Anderson & Gehrke, 1987).

Many dams in the West have been placed in areas of sparse vegetation with erodible soils and periodic rain of six inches a day or twenty inches in a month. These conditions greatly exacerbate the problem with silting behind the dams. All reservoirs eventually silt up but, situated in a location with a lot of hard rock and vegetation, a dam may be useful for a thousand years.

Water contamination. The major aquifers located beneath the eastern part of the country are not threatened with depletion but with toxic chemical contamination. More than half of the population of the nation obtains its drinking water from individual wells or public groundwater systems and contaminated groundwater is very difficult to restore. The chemicals used for agricultural pest and weed control, nitrate in fertilizer, mine tailings and garbage are polluting these regional aquifers and wells (Gottlieb, 1988; Powledge, 1987).

Serious damage to fish, birds and other wildlife has been linked to agricultural drainage water contaminated with selenium and other toxic chemicals. It is estimated that more selinium and other potentially lethal chemicals have been washed out of the soil, through irrigation, in the past few decades than would have occurred through natural rainfall in centuries (Postel, 1989).

Conclusions

The primary solution to the problem of water conservation in the West would appear to be pricing reforms and considering changing the water laws to allow water to be marketed. In this manner, the farmer could sell the water conserved to the highest offerer. As Reisner (1990:59) points out:

> . . . the transfer of water rights shows great promise as a means of achieving several important goals at once; supplying water-short urban areas *while* alleviating the drainage and salinity crisis *while* reducing surplus crop payments *while* promoting ecological health -all at a reasonable cost *without* new dams.

What constitutes *beneficial use* needs to be carefully thought out. Whether tourism and recreation should be included is an important issue. The provisions defining beneficial use are frequently vague and difficult to enforce.

Water could also be conserved by selecting crops with the most minimal water consumption needs. For example, lettuce and other vegetables tend to be high value crops and low-water intensive, but in 1980 they were under five percent of Arizona's irrigated cropland while alfalfa, a high-water consumptive and low value crop, constituted nearly 16 percent of irrigated acreage (Welsh, 1985). In California, more than half of the acre feet used in agriculture is irrigated pasture and alfalfa--used to feed cattle, not people (Powledge, 1982). Other suggestions include: focusing more on native plants which require little water, using drip irrigation systems, low-energy precision application and laser leveling, lining ditches, and reusing water.

Reusing water has great potential. One irrigation district in the Tucson area estimates that by substituting 40 percent of its pump water with waste water, they have gained over $4 million

(Welsh, 1985). Further, waste water contains nutrients such as phosphates and nitrogen which are excellent for plants. Waste water would be particularly useful on non-edible crops such as alfalfa and cotton. However, improvements in water efficiency are not likely as long as Western farmers obtain water at highly subsidized rates.

Finally, conserving water is going to require taking some of the land in the West out of agricultural production. For example, many people in Arizona argue agricultural development is not practical with water in scarce supply; thus, development should be in terms of population and industry. Others would argue any kind of development in the West should not be encouraged.

7

Farmer Participation in Irrigation Management: Limited Roles in Pakistan

*W. Randall Ireson**

Introduction

Within the last decade, most social scientists working with irrigation systems have come to agree that farmer participation in management greatly enhances the efficient operation of the system. While this conclusion is gradually being accepted in principle by some system managers, as well as by bilateral and multilateral donor agencies, there remains a significant gap between this acceptance in principle and the effective inclusion of farmer input in the management of an irrigation system. This gap is a result of at least three factors: the unwillingness of entrenched irrigation bureaucracies to share their power with water users; the lack of knowledge of managers (usually irrigation engineers) regarding the kinds of participation which can be elicited from farmers; and the lack of awareness of important social obstacles which inhibit effective mobilization of water users for participation in irrigation management. This paper describes the nature of farmer participation in irrigation management as it occurs in Sind Province of Pakistan, comparing traditional areas of involvement

*This paper derives from the author's work in Pakistan with the Command Water Management Project. The ideas expressed herein are solely those of the author and do not necessarily represent the position of the Government of Pakistan, Associates in Rural Development, Inc., Louis Berger International, Inc., or the U.S. Agency for International Development.

with expanded participation which is being encouraged through the Command Water Management Project (CWMP). Obstacles to increased farmer involvement are discussed, and an analysis of the potential for the CWMP to overcome these obstacles is presented.

The management of irrigation systems occurs at different levels and involves different tasks. Freeman and Lowdermilk (1985) identify central organization, local command area, and farm levels of water management. Easter (1986) distinguishes between reservoir, canal system, and farming system management, while Uphoff (1986a) more generally numbers the levels upward beginning with the field channel. A discussion of farmer participation must, therefore, identify the level(s) in the system at which participation occurs or is desired. Frequently farmers are assigned the lowest (field) level to manage themselves, but have little input or influence over the operation of the system which supplies water to the field channel. Such an arrangement has obvious implications for the relationship between the farmers and technicians managing the higher reaches of the system. On the other hand, farmers can be involved in the operation of higher levels of the system if an appropriate structure of representation is developed. Even very large irrigation networks can include farmer input into decisions regarding water scheduling and maintenance (Bautista, 1987), though it is safe to say that such participation does not occur without considerable effort, planning, and support from government agencies and officials.

Any irrigation system must also accomplish a variety of tasks, and it is possible to include or exclude farmers or water users from participation in these. Coward (1985) identifies five fundamental tasks: water acquisition, water allocation, system maintenance, resource mobilization, and conflict management. Uphoff and his colleagues (Uphoff, 1986a; Uphoff et al., 1985) expand this list to a 4 by 4 by 4 (64 cell) matrix plotting the intersection of organizational, control structure, and water use activities. In this scheme, organizational activities are those of decision making, resource mobilization, communication, and conflict management. Design, construction, operation and maintenance, are control structure tasks. And water use is classified into acquisition, allocation, distribution and drainage activities. The classification is somewhat cumbersome, but provides a rather precise framework for identifying the areas in which water users may participate in irrigation management, for example, how they are involved in resource mobilization for construction of water distribution structures. The following discussion of farmer participation in irrigation will, therefore, make use of these analytical frameworks.

The Indus River Irrigation System in Pakistan

The Indus River irrigation network has been called the largest irrigation system in the world, though, in fact, it is more a coordination of several very large systems all drawing water from the Indus and its tributaries. In all, over 13 million hectares in Pakistan are irrigated from this source. The modern development of the system began during the 1930's with British construction of the Sukkur barrage, together with a network of main canals feeding distributaries, minor tertiary canals, and eventually watercourses. Kotri and Gudu barrages were added in 1958 and 1966, and Tarbella Dam was completed in 1975. Additionally, there are numerous smaller barrages in Punjab on the tributaries of the Indus. The system was initially designed to provide supplemental irrigation to what was basically an arid region agricultural regime, but in the ensuing decades the demand on and capacity of the network has expanded to the point where virtually all farming in the Punjab and Sind is dependent on irrigation, and cropping intensities of 120 to 150 percent are common in the irrigated areas.

Water has historically been allocated through a constant discharge system. In principle, all waterways carry a constant flow at all times (subject to some seasonal variation), which is divided in fixed proportion at each junction as one proceeds to lower levels of the canal network. The Provincial Irrigation Departments (PIDs) are responsible for maintaining the design flows from the diversion barrage through the main, distributary and minor canals, and delivering a constant discharge through each final turnout into the watercourses which supply farmers' fields. Below the turnout, or *mogha*, water is allocated to individual farmers in a seven or ten day rotation known as *warabandi*, where each farmer receives the entire discharge of the watercourse for a number of hours dependent on the proportion of land he owns in the overall watercourse command area. In most cases, the PID has calculated and sanctioned the *warabandi* schedule, but in practice, the responsibility of the department ends at the *mogha* and farmers are usually left to manage the water distribution themselves.

As the overall irrigation network was enlarged and improved, the discharges delivered to the various canals have increased. For example, distributary canals branching from the Rohri canal (at 400 kilometers, the second longest of seven canals originating at the Sukkur Barrage and supplying irrigation to about 500,000 hectares in Sind) now typically carry from 140 to 170 percent of their original design capacity. As water allocations to the individual watercourses have not been revised since the early 1960's, it is a relatively easy matter for the Irrigation Department to maintain at least the sanctioned discharges at almost all turnouts, even while some watercourses may receive two or three times their sanctioned flow.

The Provincial Irrigation Departments are powerful bureaucracies in their own right. Overall allocation of Indus basin waters to the various barrages in Pakistan is made by the Indus Basin Commission. But, water allocations to the various canals, maintenance and rehabilitation activities (and budgets), monitoring of canal conditions and discharge on a continuous basis are all activities carried out by the officials of the provincial departments. Virtually all activities related to water use, control structures and organization down to, but not including, the turnout, are the exclusive domain of the PID. A few minor exceptions to this generalization will be discussed below. As with most Pakistani bureaucracies, the irrigation departments are administered top-down, with directives and operating procedures issued from above to be implemented by divisional and field level personnel. Within this framework, the particular irrigation needs of farmers, whether in a district or on a watercourse, play a very small role in the overall allocation of irrigation water. The sanctioned discharges were originally computed, based on a given water duty (around 4.25 ha per lps in northern Sind, but varying from region to region), and the objective of the PID is to supply the required water on a continuous basis. Other than usually scheduling annual canal closures for maintenance at times of low crop water requirement, few official concessions are made to the agricultural system. However, there are several extra-legal ways in which some farmers interact with the department to modify their water allocations; these will also be discussed below.

Watercourse Management

Below the *mogha*, water management is almost entirely the province of the farmers. Each watercourse supplies a command area which may range from less than 80 to nearly 400 ha and which typically is managed independently from any other watercourse command areas. As noted previously, water is usually distributed among different landowners by means of the *warabandi* system, which may or may not be officially sanctioned by the PID. Unofficial *warabandi* must be managed entirely by the farmers on a watercourse, but even in official *warabandi*, the PID seldom takes any active role in the distribution of water. Conflicts must be resolved by the farmers themselves, or by reliance on powerful local figures. Despite some studies indicating a high level of conflict over water in Punjabi villages (Merrey, 1986a; 1986b), other experiences suggest that water distribution proceeds routinely (which is not to say equitably) in much of the Sind, though there are undoubtedly tensions which occasionally lead to overt conflict.

One possible reason for this apparent lack of conflict is the pattern of land ownership in Sind and its relationship to social stratification. Land-holding is highly concentrated, and around two-thirds of all farmers are tenants. It is not uncommon for over half the land on a watercourse to be owned by one or two families, in contrast to the smaller and more equal-sized holdings in the Punjab. In this situation, the influence of the major landowner(s) is often enough to enforce a *warabandi*, even one which gives him proportionally more water than the other farmers. Furthermore, allocation of irrigation water within the lands of a single large landowner (*zamindar*) is not dependent on a scheduled rotation but on the *zamindar's* own priorities. He can, therefore, supply his various tenants according to his perception of their needs and desires. (The male pronoun is used throughout to reflect the exclusion of women from any public role in Pakistani agricultural activities.)

Watercourse maintenance and repair of damage is also the responsibility of farmers in the command area. This task is not always performed, or performed well. Until recent foreign aid projects focusing on water management, it has received no attention from the Irrigation Department or other government agencies.

The only other common area of farmer involvement regards the use of tubewells. Private tubewells are particularly common in Punjab, though less so in Sind. Additionally, a large number of government tubewells have been installed in both provinces under the Salinity Control and Reclamation Project (SCARP). SCARP wells and some private wells discharge into the common watercourses, and provide a significant addition to the overall irrigation supply. Private wells, which can be operated at any time the owner wishes, provide a degree of flexibility to irrigation scheduling which is otherwise unavailable, and other farmers are willing to pay tubewell owners for this water (Wattenburger,

1986). SCARP wells are, in theory, operated under a more rigid eight-hours-per-day limitation, but some farmers have succeeded in extending the operating hours of a well on their watercourse.

Examples of Watercourse Organization

A description of farmer behavior regarding water management activities on several watercourses will clarify the scope and variation of their participation. These examples are drawn from conversations with farmers on numerous watercourses in upper Sind, and while no statistical generalizations are possible, the descriptions are representative of the range of farmer behavior encountered in the region.

Watercourse A provides an example of longstanding farmer cooperation in local operation and maintenance activities. The watercourse command area is 328 ha, farmed by around 40 growers. Five have more than 30 ha each. For forty years, one farmer, not one of the five largest landowners, has been the leader for watercourse maintenance. He is the only literate farmer on the watercourse. Several times a year he would tell people a day in advance to send workers to clean the channel. There was a minimum labor requirement of one worker for every *pahr* (3 hours of water per week) a farmer was allocated, though commonly several men from a family would come. The main watercourse could be cleaned and strengthened in one day, after which farmers separately maintained their own branches. If a farmer did not come to work, others would speak to him about it, and, as a last resort, impose a fine. However, no fines have been levied in at least the last five years. About a year ago, the watercourse received partial lining through a government aid program. At this time a Water Users Association (WUA) was formed, fulfilling a legal requirement to qualify for the assistance. The traditional leader became the chairman of the WUA, and has continued to organize maintenance activities as before. Labor for the lining work was provided by the farmers, either on their own, by hiring laborers, or by directing their sharecroppers to do their share.

Water is distributed according to a *warabandi*, but occasionally farmers trade water on an hour-for-hour basis. At times when the discharge to the watercourse has decreased, the leader/chairman has gone to the PID, sometimes with other farmers to complain. Reportedly, they have usually received satisfactory responses.

A contrasting example is provided by Watercourse B. This watercourse was also partly lined through the same government program about a year ago. However, only two farmers, each owning about 25 hectares of land, hired laborers and sent their sharecroppers to carry out the construction. One of these farmers is the chairman of the WUA, while the other *refused to join* because he was not totally satisfied with the decision regarding which portion of the watercourse would be lined. According to the Sind Water Users Association Ordinance, however, if enough landowners on a watercourse agree to form a WUA, all the landowners automatically become members of the association. The farmer is unaware of this provision and his corresponding automatic membership; but since the WUA, in this case, is only a paper organization formed to fulfill a legal requirement, his membership or lack of same is irrelevant.

Maintenance prior to the improvement was performed very irregularly--only when there were adverse conditions affecting a farmer. Then, that farmer would work to repair the damage which affected him directly. There was no cooperation between the farmers; because the two *zamindars* mentioned had the largest holdings, they ended up doing most of the maintenance. Now, they simply hire laborers or require their sharecroppers to do the necessary maintenance. The other farmers on the watercourse, numbering over 100, mostly have between one and four hectares of land and do not participate in any maintenance work.

Water distribution on Watercourse B is also by *warabandi*, supervised in part by a field official from the PID. The allocation is about 35 minutes per hectare, but the smallest farms get a minimum allocation of 30 minutes. Farmers occasionally trade irrigation times by mutual agreement on an hour-for-hour basis. A government tubewell also supplies water to the farmers, but it has been out of commission for six months. The farmers paid the operator *rupees* 1200 (about US $68) to repair it, but he has not done the work.

Watercourse C provides a third example of farmer involvement in the irrigation system. The command area is 192 hectares, divided among 40 landowners. The main *zamindar* owns 16 hectares and leases another 25. He has been the organizer of maintenance activities for many years, and became the chairman of the WUA when it was formed prior to lining part of the watercourse. Traditionally, he would call farmers to clean the watercourse from four to six times a year; small farmers would come themselves and larger farmers would send their sharecroppers. The work on the main and larger branch channels could be completed in a day, after which farmers would maintain their smaller field channels. After formation of the Water Users Association, the lined section needs cleaning only about twice a year, but the unlined sections are maintained as before. During the construction of the lining, all the farmers contributed labor, though some had to be persuaded by the chairman. The chairman also provided meals for the workers during the three weeks construction was in process--a rare case where food was provided for any cooperative labor.

Also, in contrast to most watercourses, the chairman calls meetings of the farmers if there are any problems, e.g., if the government tubewell breaks down, or if there are problems in dealing with the Irrigation Department. While SCARP is supposed to maintain tubewells, in reality farmers must secure their own repairs. Thus, the chairman assesses the farmers on a per-hectare basis to hire a mechanic from a nearby town, and pay for his transportation, tea, and spare parts. Through judicious bribes to PID officials, the farmers can also double the water supplied to their watercourse. These payments are also secured through a per-hectare assessment by the chairman. Internal water distribution is by *warabandi* at the rate of 52 minute/ha.

A final example describes the common situation where a dominant *zamindar* simply controls the activities of other farmers regarding irrigation. The chairman of the WUA on Watercourse D owns about 28 of the 100 hectares in the command area, but also owns land on 10 other nearby watercourses. Maintenance is performed at his initiative, perhaps four to six times a year. At these times, he calls the farmers to work and assigns a length of watercourse to clean that is proportionate to their land area. His influence is such that he can ensure everyone does his share within a grace period of a few days. Water distribution is by *warabandi*, but as the chairman has a 48-hour share for his own land, he is relatively free to allocate among his different fields and tenants. The orchards which he cultivates himself (with hired laborers) receive a greater share of water than the wheat or cotton area of his sharecroppers. Before the watercourse was lined, the chairman unilaterally decided to approach the relevant government officials for assistance, obtained the signatures from the other farmers which were necessary to organize a Water Users Association, and then required his own sharecroppers to provide all the labor needed for construction. He did not expect, nor did he receive, any assistance from the other farmers on the watercourse.

There are also cases where all land on a watercourse is owned by a single family. The family then takes responsibility for operation and maintenance of the watercourse as an internal matter, and typically allocates the work and water among their sharecroppers. When necessary to secure government assistance for watercourse improvement, a Water Users Association, composed of the different relatives whose names are listed on land ownership records, is registered. Local political influence can also be used to secure a greater-than-allotted share of canal water for the watercourse.

Analyzing the Pattern of Participation

The above examples give some idea of the range of farmer involvement in irrigation management in Sind. Considering them in the context of Uphoff's (1986a) classifications, we note that Sindhi farmers participate in irrigation activities in the following ways:

(1) water acquisition, but only to the extent that they can manipulate the discharge from the minor canal to their watercourse, or the timing of tubewell operation;

(2) water distribution, in the operation of a *warabandi* on the watercourse. On watercourses without an official *warabandi* based on land area, farmers must also make the allocation decisions as well;

(3) control structure maintenance, regarding the watercourse, the tubewell, if there is one, and occasionally sections of the minor canal;

(4) construction of control structures, but only within the framework of the government watercourse lining program;

(5) resource mobilization for the maintenance and construction activities noted, and to raise money to bribe irrigation officials; and

(6) conflict management, but usually confined only to issues of water distribution and labor mobilization for maintenance.

The manner in which these activities are carried out varies from one watercourse to another, primarily in terms of internal decision-making and the means available for communication between farmers and the PID. Methods of water acquisition and distribution illustrate some of the patterns observed. If some of the land on a watercourse is owned by a very large landholder, he may arrange for the mogha to be enlarged (usually by removing several bricks from the sill of the outlet). Such tampering may not be challenged by irrigation officials or other farmers because of his influence in local affairs. This influence may take such forms as influencing the local police to bring specious complaints against a person, causing an official to be transferred to an undesirable post, or, in extreme cases, arranging for local bandits (*dacoits*) to kidnap a person. These sanctions are not often resorted to, but the potential for their exercise seems to inhibit the willingness of officials to enforce regulations against the behavior of such *zamindars*.

If no *zamindar* of this status is found on a watercourse, but other landowners with significant holdings (e.g., over forty to eighty hectares) exist, they are often in a position to intercede with the PID officials. In exchange for the payment of several thousand *rupees*, the officials will ignore extra-legal modifications to the outlet for a season. This relationship is fairly routinized, as the irrigation officials may replace the bricks in the *mogha* at the beginning of a growing season under the guise of correcting irregularities. But, upon receipt of the appropriate payment from the farmers on the watercourse, the obstructing bricks are again removed.

Watercourses with no large landowners are in a comparatively disadvantageous position. With no one of sufficient social status or familiarity with the bureaucracy to intercede with PID officials, efforts to increase the irrigation discharge to their fields are subject to the prior acquisition of water by watercourses with more influence. Similarly, acquisition of extra water by farmers on watercourses in the tail reach of a minor is subject to prior appropriation by watercourses in the head and middle reaches. Even farmers who would otherwise be able to pay the Irrigation Department to overlook an enlarged outlet are unable to increase their discharge if much of the supply has already been diverted.

Water distribution among farms on a watercourse follows similar patterns. In nearly all cases, there is a formal *warabandi* in operation on the watercourse which provides the basic framework for water sharing, and generally farmers follow the schedule. Where necessary, enforcement of the rotation is provided by the larger landowners on the watercourse (or their farm managers) through their informal power and influence. This pattern seems to operate both where there are very powerful landowners, as well as locally influential *zamindars*. And if some of these choose to take a greater time share, no one is willing to make an issue of it. Infrequent water trading between farmers is also practiced on most watercourses, though in principle, it is illegal.

Farmers on watercourses which lack powerful landowners must, by necessity, rely on internal cooperation to manage a warabandi. Locally respected persons may facilitate conflict resolution and the orderly operation of the rotation, regardless of their land owning status. In some cases, the farmers distribute water with little difficulty, while in others, there may be conflict.

Patterns of resource mobilization, particularly of labor for periodic maintenance of the watercourse, also vary according to the land-holding pattern, though in a somewhat different manner. Influential *zamindars* can act as conveners as on Watercourse D (discussed earlier), and coordinate the contribution of workers from all the landowners, with self-cultivators working and larger landowners sending their sharecroppers. Alternatively, the larger landowner(s) may choose to perform the maintenance himself, either using hired laborers or his sharecroppers, and not bother to organize or involve the smaller farmers on a watercourse. Watercourse A provides an example of this behavior.

Annual cleaning of minor canals provides the only example of official interaction between the PID and farmers on a watercourse. At the time of the annual canal closure, local PID officials may contact selected *zamindars* and request them to mobilize a labor force to de-silt and clear weeds from a given section of the minor canal. However, farmers from every watercourse are not involved.

Tubewell repairs, while officially the responsibility of SCARP (not the Irrigation Department), are generally not carried out unless farmers make a payment to the tubewell operator to perform the repair or hire a mechanic from a nearby town. Landowning status and political influence does not seem to facilitate obtaining tubewell repairs from SCARP.

If we look at the structural aspects of farmer participation as described here, we note that farmers are involved in most of the 12 activities identified by Uphoff (1986a). However, except for the contact between farmers and the Irrigation Department regarding illegal modifications of the discharge at the mogha, and farmer recruitment through *zamindars* for cleaning the minors, all aspects of farmer participation noted above occur entirely within a watercourse command area and are informally organized. That is to say, there is no active interface between the farmers and the rest of the irrigation system nor are there traditional formal organizations for water management on the watercourse. This is particularly evident when we examine what Uphoff (1986a) terms *the organizational activities of decision-making and communication.* The Irrigation Department makes decisions about the operation and maintenance of the main system, and communicates them to its officials. About the only information communicated to farmers is the date of the annual canal closure. Similarly, farmers on a watercourse make decisions and communicate among themselves regarding water use and control structure activities on the water course (mostly distribution and maintenance). But generally they have no contact with the Irrigation Department except regarding *mogha* discharge.

Despite the size of the overall system and the coordination necessary to distribute water to the tens of thousands of watercourses throughout Pakistan, as far as the farmer is concerned he is only able to affect aspects *below the mogha*. The system is organized such that the irrigation departments are basically responsible for all operations from the Indus or its tributaries down to the *mogha*, but they take no responsibilities on the watercourses. For the farmers, the situation is exactly opposite. According to law, the PID can sanction a *warabandi* schedule, and the recent (early 1980s) provincial water users association ordinances also provide for the PID to require a WUA to carry out maintenance on their watercourse (Government of Pakistan, 1984). In reality, however, farmers almost always operate their *warabandi* independently of the irrigation departments, and there have, to date, been no instances of PID staff taking any role in watercourse maintenance.

A second important generalization is that a farmer's responsibilities and expenditures (in time or money) for watercourse operation and maintenance tasks are generally not proportionate to his benefits from the irrigation system. There are several factors which explain this situation. First, even in situations where farmer responsibilities for maintenance, for example, are allocated according to land area or hours of water received, the effort expended is not related to water quantity because of the high water losses between the head and tail of a watercourse (Haider, Sheng, & Tinsely, 1986) report losses of over 50 percent on some watercourses).

Furthermore, watercourse maintenance is infrequently allocated according to land ownership or water rights; more commonly either everyone works together until the job is done, or the small farmers are organized by the local *zamindar* to carry out the job. Further complicating the picture is the status of tenant farmers. Tenants in Sind (who are nearly all sharecroppers) have no rights to irrigation water, but rather receive water through their landlord, according to his allocation. Nor can Sindhi tenants be members of Water Users Associations. However, a landlord who has watercourse maintenance responsibilities will assign the task to his tenants or to his hired laborers, whose work responsibilities are not related to their access to water.

Water charges imposed by the government are minimal, and again bear no relationship to quantity of water received. For example, in the Rohri command area, SCARP tubewell charges are *rupees* 40 ($2.27) per hectare per season, regardless of the state of repair of the tubewell, its discharge, or the number of hours pumped. Similar nominal charges are assessed for canal water. These charges are collected by the Revenue Department at the same time as land taxes, thereby weakening the farmer's perception of the payment as being related to water access.

In short, the irrigation system operates almost as two separate entities. First is the system of main canals, distributaries, and minor canals, managed by the provincial irrigation departments and designed to deliver a minimum sanctioned discharge to each turnout on every minor. Second is the collection of separate watercourses, each managed in a different way by the farmers, and each receiving a particular supply of water at the turnout that must be distributed within the command area. The point of contact between the two systems is the mogha or turnout, and it is at this intersection that most interaction between farmers and the irrigation departments occurs. Because the PID has the power to determine the discharge to a watercourse, farmer action to acquire water (one of only two activities which transcend the watercourse boundary) must involve the PID. But discharges to each watercourse are fixed by regulation, and can only be changed legally if the watercourse changes, as in command area or number of hectares approved for orchards (which are allocated a greater irrigation supply). Thus, most farmer interaction with the PID regarding water acquisition must occur illegitimately in the form of bribes or political pressure to ignore excess withdrawals at a *mogha*. In fact, most interaction between the PID and farmers occurs within this extra-legal context.

Farmer-originated official requests for approval of new turnouts, or the transfer of land from one watercourse command to another, do occur, but very infrequently; the irrigation network in Sind is well-established and there is now little room for modification of the distribution pattern. Farmers also occasionally contact the PID regarding disputes over water theft both within and between watercourses, but it is unclear how responsive the PID is to their complaints. Irrigation Department mobilization of farmers for annual minor canal cleaning is about the only other area of legitimate interaction between the two sectors, and while canal cleaning benefits the farmers, they perceive it as an activity done for the PID.

This operation pattern has three important dysfunctional implications. First, because there is no effective communication between the top and bottom of the system, it is not possible to allocate and distribute water rationally on the basis of crop needs, or to modify cropping patterns on a watercourse or minor to make more efficient use of the water available at different times of the year. Because farmers are not informed what their real irrigation supply is or will be at different times during the season, they must rely on their own past experience of seasonal patterns in an attempt to plan cropping patterns (crop mix, planting dates, etc.) which will not be adversely affected by seasonal or unexpected water shortages.

Second, because the bulk of farmer-Irrigation Department interaction is extra-legal, each party tends to perceive the other as an adversary, and not to be trusted. This attitude further inhibits possibilities for cooperative management of the irrigation system, and encourages further attempts to secure an advantage, whether a greater water supply for one's own watercourse (at the expense of others), or a higher payment from farmers for overlooking their excess withdrawals.

Third, because the PID has no responsibility for irrigation management at the farm level, and is perceived by farmers as an adversary, farmers have no source of information they trust regarding better methods of water management and watercourse maintenance and repair. Partly in consequence, water conveyance and application efficiencies are extremely low (Painter, Baldwin, & Malone, 1982; Haider et al., 1986), crops frequently experience water stress, and farming yields are considerably lower than might otherwise be expected.

Increasing Farmer Participation

Several years ago, in an effort to rectify the low efficiencies in the Indus Basin irrigation network, the World Bank and the U.S. Agency for International Development (USAID) began a program with the Government of Pakistan to rehabilitate certain physical aspects of the canal system, and also to increase farmer participation in the system's operation.

This program, the Command Water Management Project (CWMP), was to combine rehabilitation and/or lining of selected minor canals, partial lining of watercourses, and the coordinated delivery of agricultural extension services, water management advice, and non water inputs (credit, fertilizer, etc.) to farmers through the Water Users Associations which were organized to participate in the lining of the watercourses. An important feature of the project was to be the provision of organizational support to the WUAs to help them deal with internal problems of cooperation, water distribution, and watercourse maintenance. Additionally, it was envisioned that information and requests from the WUAs would be accepted by the PID as a constructive input to be included in overall management of the irrigation system. A new government body, the Sub-Project Management Office (SMO) would be established, staffed with officials from both Irrigation and Agriculture Departments, but administratively responsible to a provincial policy committee rather than a particular department or ministry. The SMO would be responsible for coordinating the activities of the various departments and line agencies involved in carrying out the project in a particular area.

Such a program to increase farmer participation is in accord with recommendations for linking central- and field-level elements of irrigation systems to incorporate local knowledge and local control over water and staffing for improved water management (for example see Freeman, 1990). Using the analytical framework of the earlier discussion traditional patterns of farmer participation in water distribution, watercourse maintenance, resource mobilization and conflict management activities would be formalized within the WUA, and the relationship between the farmers and the PID would be regularized and strengthened, especially regarding water allocation decision-making. The WUAs would additionally become a vehicle for involving farmers in new cooperative activities: improved methods of water management (both along the watercourse as a whole and on individual farms), higher quality maintenance of the watercourse, access to extension information, group credit, and perhaps, group marketing of farm products. The latter three activities have little direct relationship with irrigation and if implemented, would transform the WUA into a multiple-purpose organization. However, this focus on the watercourse, while it can strengthen internal farmer cooperation, neglects a consideration of the social institutions within which farmers act. These institutions are functionally exogenous to a determination of farmer participation (Easter & Welsch, 1986), and in Pakistan, present significant obstacles to farmer participation in irrigation management on a watercourse.

The accomplishments to date of the CWMP have been modest. Watercourse lining and improvements of the distributary system are proceeding more or less on schedule; however, the involvement of WUAs in activities other than watercourse lining has occurred only in isolated cases. To some extent, the lack of progress regarding WUA development is a consequence of low numbers of staff assigned to this task, and the ease and routine of more normal civil works activities, which thereby absorb most of the energy and resources of the project. However, other reasons can also be identified which suggest that development of formal, multiple purpose WUAs may be inappropriate, and in any case will be quite difficult in the Pakistani context. Wade for example has discussed structural impediments to cooperative irrigation management in a number of papers (1982a; 1988; 1990), and concludes that a syndrome of anarchy can develop when irrigation systems are seen by farmers as being government owned, don't meet farmers' expectations regarding water supply, and cannot control free-riding behavior by irrigators.

More general social constraints to cooperation can also be identified. In two related papers, Byrnes (1985; 1986) outlines a series of conditions which affect the ability to form farmers' organizations of any kind. These conditions reflect characteristics of the farmers' organization, as well as dominant institutional patterns in the farmers' society which affect his expectations as to whether other members of an organization are likely to support it or attempt to free ride on the efforts of others. The conditions are relevant to an understanding of the potential for WUAs to stimulate increased farmer participation in irrigation and are outlined as follows (here renumbered from Byrnes, 1986:34-5):

(1) Group size: that the group's size (i.e., number of members) is appropriate relative to the specific benefit or good provided by the group to its members.

(2) Group identity: that the actual or potential members of a group (or organization) see or identify themselves as being members or potential members of the group by virtue of having one or more characteristics in common.

(3) Group structure: that the group has an adequate organizational structure in terms of leadership as well as managerial, administrative, and financial procedures.

(4) Group good: that it is impossible to exclude any group member from consuming a group-provided good or benefit if one member consumes it.

(5) Divisible good: that it is possible to divide a group good so that individual members of the group may utilize it.

(6) Individual profit: that individual or collective use of a group good by group members enables each individual to earn a minimum acceptable net return after costs and risks are taken into account.

(7) Compensatory profit: that the level of return enjoyed by a group member's use of a group good is sufficiently high to cover, not only the required level of individual profit, but also the transaction costs and loss of individual discretion involved in joining, cooperating with, and supporting the group.

(8) Organizational good: that a group good will not be available to the group's members unless they (the potential beneficiaries) organize themselves to provide it.

(9) Group sanction: that utilization of collective action to seek a common benefit or good through a group-oriented approach is sanctioned within the society in which a group is to function.

(10) Functional identity: that a group good functions in the same way for all a group's members (e.g., in an irrigation scheme, different sized farms receive the same amount of water per unit area).

These criteria are fairly straightforward and represent, if not all, at least a significant number of the criteria likely to be considered by farmers making rational decisions regarding participation in a cooperative group. We may further divide this list of ten into conditions of three types. First, there are conditions related to the structure of the farmer's organization itself (how it is organized). These include Item (1) and some aspects of Item (3). Deficiencies or problems regarding these conditions can be met by attention to the design or organization of the farmers' group. Second, some conditions are related to the actions chosen by the group and their profitability for the members. These include Items (4) through (7) and some aspects of Items (2) and (8). Problems regarding these conditions can, to some extent be met by attention to group structure, but to some extent, are also dependent on the nature of the goal selected and its characteristics. Thus, for example, an association created to purchase a tractor for use by its members may be able to divide the tractor among all members (number of hours of work), satisfying Condition (5), but excessive demand for the tractor may make it impossible for all members to get their share of time, thus violating Condition (4). Third, some conditions are embedded in the society at large and provide an institutional context in which the farmers' organization acts. These are items (9) and (10), and aspects of (2), (3), and (8).

For Water Users Associations in Pakistan, satisfying the organizational structure conditions is not overly problematic. WUA members share the common characteristic of being farmers on a particular watercourse, the number of members involved is reasonably small, and there is a legal framework for the formation and actions of the WUA, though members or officers may not be experienced in organization management or leadership.

Regarding actions of WUAs and their benefits for the members, the situation is mixed. If the watercourse receives water, all farmers have theoretical rights to an appropriate share, though there are irregularities possible in this regard. Lining a part of the channel presumably benefits all farmers to some degree, but not necessarily equally. Regarding other agricultural inputs, farmers now purchase fertilizer and, to a limited degree, receive credit through independent sources; it is not immediately evident that a WUA can substantially increase the supply of these resources, or distribute them fairly and uniformly. Whether the effort necessary to form and operate a WUA is adequately compensated by increased water supply, better distribution, and/or access to inputs or extension information cannot be determined without more farmer experience with these activities.

The main areas of difficulty regarding WUA development in Pakistan are derived from the institutional context within which individual farmers and WUAs act. Considering the conditions relevant to collective action in Pakistani society, one observes there is little precedent for farmers to trust farmer groups or organizations outside their own families. Except for some limited actions within brotherhood groups, there is little tradition for allowing group decisions to govern the use of personal resources, although some farmers trade water turns, farm labor, and animal traction on a reciprocal basis. There are no private voluntary organizations, religious organizations, or individual religious leaders who support farmer organizations. One major function of WUAs is to provide a vehicle for repayment of watercourse improvement expenses and to collect water charges--activities which are unlikely to be supported by farmers. The past history of government and informal organizations in Pakistan has been dismal regarding protection against rule-breaking or co-optation of an organization for personal gain.

Regarding the functional identity conditions, virtually all aspects are negative. In most irrigation schemes, some users consistently draw more than their alloted shares; in many credit schemes, large landowners force their tenants to apply for loans which are then collected and utilized personally by the landowner; in many production credit schemes, farmers who receive chits for purchasing fertilizer or seed from local merchants frequently exchange the chits for cash rather than using the inputs. In general, there are no effective penalties available to apply to powerful members of a group who ignore the rules.

In general, these observations describe a situation of distrust, factionalism, and lack of confidence in institutions. The acceptance of corruption as being the *way things are* inhibits efforts to create honest and responsible organizations. Farmers recognize that water access is often a function of a person's wealth and political influence, both within a watercourse and between watercourses, and believe the PID abets distribution to the privileged to the privileged. The civil unrest and lawlessness which have prevailed in Sind since 1989 further obstruct opportunities for cooperation.

Under these conditions, ordinary farmers will be reluctant to participate actively in a WUA regardless of its supposed benefits because they have little confidence that the WUA will function as designed, or that they would have any influence over its direction or behavior. And, if the benefits from a WUA are vague, or can be obtained even inefficiently through individual action, there is less reason for a farmer to take the substantial risk of joining. However, large *zamindars* or influential farmers have the experience in and ability to control or manipulate local institutions and would be supportive of forming WUAs as an additional resource for the exercise of their influence. Until now, farmers have formed and registered WUAs only to satisfy the legal requirement to qualify for watercourse lining assistance. The benefits of lining are rather universally accepted, and the effort to register and mobilize a group for the specific and time-limited task of providing labor (individually, or by hiring workers) for a 4 to 6 week construction project is usually not considered extreme. However, the history of the CWMP has, so far, been that once the watercourse improvements are complete, the WUA effectively disappears, and pre-existing forms of farmer participation in irrigation management continue unmodified.

The goal of increasing the formal involvement of farmers in the irrigation system is not being met, and future prospects are uncertain. The institutional barriers to trust, mutual cooperation, and faith in government institutions inhibit farmer willingness to approach the Irrigation Department. At the same time, the interest of PID officials in continuing the performance of their tasks in a routine manner, and in preserving their access to extra-legal remuneration inhibits their interest in incorporating farmer involvement into the system's management. To go further and develop WUAs as multiple-function organizations would require an even larger degree of farmer trust and cooperation, together with the participation of banks and other government agencies which have been notably unresponsive to small farmers in the past. Farmers' knowledge of, or prior experience with, other cooperative organizations makes them understandably skeptical of new attempts.

Discussion

At this writing, efforts to develop formal and functional WUAs are relatively new. Thus it may be premature to write off the efforts of the CWMP. However it is possible to predict the likely outcome of the project, assuming there are no changes in policy or overall level of commitment to farmer participation. As noted above, formal inclusion of farmer input to the system-wide management of the provincial irrigation networks is not likely, but farmer participation in watercourse-specific activities will certainly continue. With proper support, these watercourse level activities can be carried out in a more regularized manner, which may improve the quality of the work, and at the same time increase farmer awareness of participation in the overall irrigation system. Many farmers on watercourses are already involved in maintenance and water distribution activities; if improvements in quality and equity of these activities occurs, along with formalization of the watercourse organization, some increased confidence and trust in the institution may develop.

The CWMP can certainly provide information to farmers regarding more efficient practices of water management at the farm level, as well as information which allows them to allocate limited irrigation resources better among their different crops. Of itself, this activity has no consequences regarding institutionalized farmer participation in the operation of the overall system, but could have an impact on overall agricultural productivity. Equally important, if this information were consistently provided through WUAs, rather than to individual farmers or as general extension information, there would be a signal to farmers that the government is supportive of WUAs as an institution, and of the farmers who are members.

It is likely that a few exceptional WUAs, perhaps because of a history of greater village cooperation or because of good leadership, will engage in some of the additional non-water related activities seen by the project designers as incentives for farmers to join an association. But, the number of groups forming credit cooperatives, or jointly purchasing farm equipment, for example, are likely to be very small.

In general, the division between Irrigation Department management of the upper system and farmer management of the watercourse command areas is unlikely to change without the express commitment of the PID to the goal of incorporating farmer input. Project emphasis on farmer cooperation alone is not adequate without similar attention being paid to the development of PID policies and staff attitudes regarding the incorporation of farmer participation in overall system operations. There are some indications that the CWMP in Sind may begin to address these issues and consciously attempt to bring the PID and WUAs together, but it is too early to predict the outcome. Specific areas where farmer participation could be well utilized need to be identified. The PID would have to perceive the farmer involvement as assisting their efforts and/or reducing their workload, and at the same time the farmers or WUAs would have to perceive their own input as giving them a genuine and legitimate influence on some aspects of decision-making in the irrigation system.

Certainly, the problems regarding irrigation management in Pakistan are not unique. Nor is it true that Pakistani farmers are uninvolved in irrigation management. However, the extent of their involvement is narrowly constrained to the watercourse command area, and to activities primarily oriented toward water distribution and local ditch maintenance. This limitation appears, in large degree, due to a combination of factors: the organization of the provincial Irrigation Departments which provide no regular avenue for farmer access; the generalized distrust by farmers of many government agencies and officials, reinforced by extra-legal actions on the part of irrigation officials; and the lack of well developed traditional models for equitable intra-village cooperation in Pakistani society.

It is sometimes taken almost as an article of faith that increased farmer participation will improve irrigation system performance. Certainly there is plenty of evidence supporting this contention, but one ought to ask about the range of participation being encouraged before assuming a particular project will have this result. Organizing farmers to perform local maintenance to more rigorous standards, collecting the funds due the government for system improvements, and modifying internal watercourse irrigation schedules which have been in operation for years, does not give farmers any greater stake in the overall system. Rather, they are required to contribute more while remaining excluded from any meaningful decision-making role. One could even argue that this type of increased government attention and advice regarding water management on the watercourse has the effect of reducing farmers' ability to make decisions at the local level.

Generalizing beyond the Pakistani case, it is suggested that where poor irrigation system performance, or low levels of farmer participation are issues, farmer participation should not be blamed for system performance without consideration of the structural factors which may promote one or the other. Where social norms and institutions do not support, or actively inhibit farmer cooperation, and where the organization and behavior of an irrigation bureaucracy formally excludes water users from meaningful decision-making roles, one should be cautious before proposing stronger water user organizations as a panacea for the performance problem. Certainly, many of the problems require social, rather than engineering solutions, but the proposed solutions need to reflect the complex nature of the social and political relationships involved.

8

Farmer Participation in Irrigation Management: The Philippine Experience

Benjamin U. Bagadion

Introduction

In the Philippines, irrigation systems are generally categorized into three types: national, communal, and private. National systems are owned by the national government; communal systems are owned by groups of farmers; private systems are owned by individual farmers. The National Irrigation Administration (NIA) is the government agency charged with irrigation development in the Philippines. It plans, constructs, operates and maintains the national systems, extends financial and technical assistance to groups of farmers in the planning, construction, and management of communal systems, and when requested, renders technical advice to private irrigation systems. As of 1986, the irrigation service areas were approximately 595,000 hectares for national systems, 640,000 hectares for communals, and 220,000 hectares for private systems.

The National Irrigation Administration was established in 1964 as a government corporate agency, under an act of Congress to accelerate irrigation development for improving the state of domestic agriculture and solving the annual rice shortage. At that time the average paddy yield of 1.7 tons per hectare in the Philippines was among the lowest in the world. The country had been importing rice as far back as 1885, often as much as ten percent of its annual rice needs. In 1964, paddy production was short of requirements by about 400,000 tons exclusive of needed buffer stocks.

The focus of NIA activities was the planning and construction of new irrigation projects to expand the country's irrigated area as fast as possible to meet the requirements of accelerated rice production. Operation and maintenance (O&M) of irrigation systems was given only routine attention. By the early 1970's, NIA's top management was becoming seriously concerned about problems in operation and maintenance. The irrigated areas of its projects were far short of the expectations of planners, and poor collections of irrigation fees were far below operation and maintenance expenses. By the mid-1970's, NIA's top management was beginning to feel that water users associations were a crucial element in effective O&M. This started a series of pilot projects designed to maximize farmer participation as an approach to organize strong irrigation associations. The commitment to find ways to develop such organizations through farmer participation started a chain of activities that resulted in changes not only in O&M activities, but also in NIA's approach to construction. This paper reviews the evolution of farmer participation in NIA's irrigation development strategy in the Philippines.

Creation of the National Irrigation Administration: Prelude to Change

Initialy, the focus of irrigation development was on engineering and construction. It was in these activities that engineers proved their worth and gained recognition and promotion for their accomplishments. The design and construction of the physical system was exciting, but operation and maintenance was considered prosaic. Talented engineers went into design and construction, while those less endowed were assigned to O&M. Yet, it was O&M that presented principal problems and strategic challenges.

The main long-term problem of NIA was the poor performance and lack of viability of the irrigation systems it managed. In the national systems, only about 80% of the service area was irrigated during the wet season, and about 30% during the dry season.

Farmers often complained of unsatisfactory service, distribution of water was generally inequitable, production was below expectations and irrigation fee collections were poor and totalled less than O&M expenses. In the communals, farmers were becoming more dependent on government and losing their initiative for self-reliant operation and maintenance of their irrigation systems. The communal irrigation associations being organized were weak and ineffectual, and the irrigated areas of communal systems tended to decrease with time. Yet, to meet the increasing rice needs of the country, NIA was being asked to accelerate the construction of new irrigation systems. Unless remedies were found, the O&M situation of irrigation systems of NIA was bound to get worse.

In the early 1960's, before the establishment of NIA, a program was started in the Irrigation Division where irrigation officials and field engineers were sent on observation tours of irrigation systems in Taiwan and the United States. The program was carried over to the early years of NIA. As a result, some features of the irrigation systems in Taiwan became models for study by many NIA engineers. The Taiwanese design of rotational irrigation became a subject of much discussion and debate in NIA. To the more perceptive observers, the most significant feature of the Taiwan systems was that they were being managed by farmers' irrigation associations that extended to the grass roots level under the supervision of the government. Those who also went to observe irrigation systems in the United States noted that the systems were being managed by farmer organizations in the form of irrigation districts under government regulation. On the other hand, the Philippine national systems were being run by the central government with personnel responsible not to the farmers, but to the central government agency. The system personnel received their pay whether irrigation service was good or indifferent, and whether or not farmers paid the irrigation fees. Farmers were not organized for effective representation in the operation of the systems and thus had no participation in the important decision-making processes. These comparisons raised the question that the root cause of the problems of Philippine irrigation was in fact the lack of farmer participation.

It was not that the government was against farmers participation. The fact that basic government irrigation legislation required repayment of O&M and construction costs by water users in national systems indicated government desire for farmer participation. This participation was in the bearing of costs, and not in management and decision-making. To NIA, such active participation was still in the realm of fantasy. The communal systems that NIA assisted were just as problematic as the national systems. Farmers in these systems had become dependent on government. Systems could not be adequately maintained since farmers sensed they could get all the construction and most of the O&M free, under the *pork barrel* system of congress.

During the early years of its existence, NIA faced a blank wall in irrigation systems O&M. Although a corporation, according to its charter, it was operating more like an ordinary government bureau. Its funds for general administration and O&M were being appropriated by the national government, and its irrigation fee collections were being remitted to the national government. The appropriations were considered as yearly advances to be repaid by remittance of irrigation fees. Every year the percentage of annual collection was so low it was always less than the annual advance. Each year the O&M financial deficit increased.

A solution was attempted in 1967. With the approval of the President of the Philippines, NIA increased irrigation fee rates from $12.00 per hectare annually to $25.00 per hectare for the wet season crop, and $35.00 per hectare for the dry season crop. It was estimated that the income from the fees would increase, enabling more funding for improving O&M services. The result, however, was disappointing. Although total collections increased in amount, the annual O&M deficit remained, because annual expenses increased: A greater effort was put into improving O&M to justify the increased irrigation fee rates. Moreover, in percentage relative to annual collectibles, the collections decreased from 59 percent to 27 percent.

The half decade that followed was focused on the Upper Pampanga River Project (UPRP), the first multi-purpose project of NIA involving irrigation, power generation, flood control, and watershed management. It was planned to irrigate ultimately about 100,000 hectares, and was funded by a loan from the World Bank. Its main structure was a 107-meter high earth-filled dam on the Pantabangan River behind which would be a 7,000 hectare storage reservoir. It was the first irrigation project of such magnitude in the Philippines, and a major block of NIA's best engineers was assigned to the project to ensure completion in accordance with World Bank agreement. The subsequent successful completion of the UPRP ahead of schedule demonstrated NIA's engineering and construction capabilities. Meanwhile, the demand for increasing rice production continued, and a series of foreign-assisted irrigation projects followed with financing mostly from the World Bank and the Asian Development Bank. In all of those projects the emphasis was on completion in accordance with engineering plans, specifications, and the time schedule agreed upon with the financing institutions. To meet these requirements, the practice of having farmers construct the farm level facilities in previous national projects was discarded. NIA had to construct the farm ditches, otherwise, the schedules agreed upon with the foreign financing institutions could not be met.

The new practice, however, had adverse effects when the projects went into operation and maintenance. Farmers complained that the location of the turnouts and the farm ditches were not responsive to their needs. Consequently, they established what NIA personnel termed as *illegal turnouts*, constructed their own farm ditches, and erased those constructed by NIA that did not fit their needs. Later, a study made in a part of the Upper Pampanga Irrigation System by researchers of NIA and the International Rice Research Institute (IRRI) indicated that about 40 percent of the area was being irrigated by *illegal* turnouts. This raised the question of whether NIA personnel still had effective control of that part of the system. Farmers were participating, but it was disorganized and conflicting in a way that NIA staff could not understand.

The Threshold of Change

In 1974, NIA amended its charter. It secured Presidential Decree No. 552 (PD 552) which empowered NIA to supervise communal irrigation systems constructed or improved with government funds and recover the funds it spends for the construction and rehabilitation of such systems. It also empowered NIA to delegate partial or full management of national systems to duly organized irrigation associations and to add to its operating capital all amounts collected, such as irrigation fees, equipment rentals, drainage fees and administrative charges.

PD 552 was the starting point of change in NIA. It mandated that the funds for general administration and O&M of national irrigation systems, including those for investigation, survey, and new project feasibility studies, be included in the annual general appropriations act of the national government. With this provision, the annual appropriation to support NIA operations ceased to be an advance fund to be repaid by NIA. Since NIA collections could be added to operating capital, sufficient funding accumulated for adequate O&M. The understanding with government budgetary authorities was to gradually reduce the government annual appropriations for O&M of national systems until phase-out. After five years, NIA was expected to fully fund O&M from its own resources. The amended charter clarified and reinforced NIA's role in communal irrigation development. The original charter transferred and assigned to NIA the functions and duties of the Irrigation Division of the Bureau of Public Works. The role of the Irrigation Division was clear in regard to the construction of communals, but it was ambiguous in regard to activities after construction. PD 552 added to the functions of NIA the supervision of operation, maintenance and repair of communal irrigation systems constructed or improved wholly or partially with government funds.

At the same time PD 552 neutralized the adverse effects of the *pork barrel* system in the communals program by directing and empowering NIA to recover funds expended for the construction and rehabilitation of communal irrigation systems. The effective use of this power, and its successful pursuit as an objective, required the development or restoration of self-reliance among the farmers through the organization of strong and viable irrigation associations.

The new charter also outlined a situation where NIA would transfer the operation and maintenance of national irrigation systems to well-organized irrigation associations. As these systems were owned by the government, specific authorization was given to NIA by the charter to delegate partial or full management of these systems to irrigation associations. Many small national systems were not financially viable under NIA for the O&M expenses; each system could not be fully covered even with 100 percent collection of irrigation fees. For a national system of about 1,000 hectares, NIA had to assign an engineer, watermasters, ditchtenders, a cashier, billing clerk, bill collector, janitor, watchmen, and laborers. Because of its corporate status, NIA compensation rates were much higher than in other governments offices, and these rates tended to increase every year. NIA employees also enjoyed more fringe benefits. On the other hand, it was not advisable to raise the rate of irrigation fees in small national systems to a level higher than that of the large systems. Furthermore, NIA had many O&M problems in the large national systems. The observation tours of some NIA officials in Taiwan and the United States pointed to the possibility of alleviating these problems through joint management of the larger systems with well-organized irrigation associations.

PD 552 also relaxed the repayment policy on construction of irrigation systems embodied in RA 3601--the original NIA charter. Under RA 3601, the level of irrigation fees should be able to finance the continuous operation of the irrigation system, and reimburse within 25 years the cost of construction. PD 552 removed the 25 year limit and made the terms of repayment dependent on government policy to introduce flexibility. Subsequently, NIA initiated, and the President of the Philippines approved, a policy that stated: the government shall bear all interests on construction costs of national and communal irrigation systems that it constructs or improves; irrigation water users shall bear the cost of operation and maintenance and pay back the cost of construction or improvement without interest within a period not exceeding 50 years, provided the level of such fees is within paying capacity and is not a disincentive; the costs to be repaid include only those for irrigation and drainage facilities within the irrigated areas of the system, and not the costs of roads, flood control, reforestation and power generation.

For a successful communal irrigation development program under its amended charter, NIA needed strong, viable irrigation associations that could sustain adequate system operation and maintenance and repay construction costs without interest within a period of less than fifty years. To NIA this meant three things: (1) a well defined legal basis for the organization and recognition of such associations so that they could transact business with the government and other entities and obtain rights to the use of water; (2) a feeling of ownership on the part of the farmers in communal systems which NIA would improve and expand, as well as new communal systems it would construct; and (3) an appropriate process of organizing farmers or improving existing organizations into strong and viable irrigation associations.

In regard to the first, the legal basis was already in place. A process for registering the associations with the government had long been established. The authority of an association over its irrigation system was clear and well defined and its registration was with a neutral government agency to promote the independence of the association. Moreover, the rights to the use of water for irrigation were granted in the name of the association. These were strengthened in 1975 when NIA staff had a leading role in the framing of a new Philippine Water Code that repealed the Irrigation Act of 1912.

In the new code, water rights for irrigation were made appurtenant to the grantee, and no longer to the land, as previously provided in the Irrigation Law of 1912. In the communal systems the irrigation association was the water rights grantee and thus, under the new code, attained full power to allocate and distribute the irrigation water in the most equitable and productive manner. It strengthened the function of the association from that of administrator of the water rights appurtenant to each parcel of land, to owner of the water rights with power to allocate and distribute the water as it may decide. The same would be applicable to irrigation associations in national systems to whom NIA would delegate authority to operate and maintain the whole or a part of an irrigation system. The new code encouraged granting water rights to irrigation associations instead of individuals. It provided that no water permit was to be granted to an individual when his water requirement could be supplied through an irrigation association. To promote better water usage, the merger of irrigation associations was encouraged. Groundwater was declared as belonging to the State and no longer to the owner of the land above. This facilitated the granting of water rights to irrigation associations from groundwater sources.

A basis for developing a feeling of farmer ownership of the irrigation system was laid out by the amended charter in a provision authorizing NIA to have farmers repay some of the construction costs of the communal system. This had been suggested in previous research which indicated that Philippine indigenous irrigation systems owned by farmer groups continued to be operated and sustained by the farmers. The researchers further suggested that investment of time and labor by the farmers in the construction of the system developed commitment to its sustained maintenance. This observation prompted NIA to require a pledge from the farmers to contribute, during the construction, ten percent of the costs of construction or improvement in the form of labor, cash or materials.

Unfortunately, NIA did not have an effective process for organizing irrigation associations. Previous efforts did not produce the desired results. On the other hand it was being pressed by the government to accelerate the development of more irrigation projects. Consequently, in 1975, NIA hired the Farm Systems Development Corporation (FSDC) to organize irrigation associations for NIA.

The FSDC was a government corporate agency which grew from a pump irrigation program launched with assistance of NIA. In that program, small groups of farmers were formed into associations to operate and maintain low lift pump systems for irrigating areas of 50 to 200 hectares. In 1975, the program was transformed by presidential decree to FSDC, a government corporation with the Secretary of Public Works, who was concurrently NIA Administrator, as Chairman of the Board of Directors. As NIA was instrumental in setting up the program which became FSDC and as the NIA Administrator was the Chairman of the FSDC Board of Directors, NIA decided to hire FSDC which had an ongoing farmer organization program to undertake the organization of irrigation associations for NIA's communal irrigation program. It was expected that the arrangements with FSDC would answer NIA's need for strong, viable irrigation associations. However, this did not happen.

Under the arrangements, NIA did the technical work and FSDC organized the farmers. The idea was that these were two independent tasks to be carried out by two separate organizations. Coordinating committees with representatives from NIA and FSDC were established at the national, regional, and provincial levels, but FSDC staff had their own programs to attend to. The work with NIA was only a part of their many activities. Despite regular meetings, the needed coordination in the field could not be attained and many problems on both sides could not be resolved. The NIA objective of having irrigation associations that could be capable of effective operation and maintenance was not being realized.

Faced with time-consuming coordination problems and uncertain about the effectiveness of the arrangements with FSDC, NIA's top management decided to start a program for integrating both technical and institutional aspects of irrigation under NIA. Furthermore, it was felt by some that irrigation institutional development was of strategic importance to NIA, and that it was unwise for NIA to depend on another agency for this on the long-term. In considering how to start the program, NIA's top management used the results of research on Philippine indigenous systems which showed that:

(1) The systems were constructed by farmers with very little, or no help from the government. The systems were small-scale, with earth canals and temporary diversion weirs of logs, rocks, and brush.

(2) The systems had been operated and maintained effectively for many years by the farmers. They often had rules for equitable allocation and distribution of water, for maintaining canals and repairing the temporary weir, for penalizing violators, and for settling conflicts.

(3) In the best of these systems, the associations were strong and based at the grass roots level. Leadership was dedicated and knowledgeable about the irrigation system. Each member knew his obligation, did the work expected of him under the rules of the association, and was penalized for failure to do so.

(4) In contrast to the irrigation systems constructed solely by the government, the farmers who constructed their own irrigation systems continued to maintain the systems.

It was obvious to NIA that farmer resources alone were inadequate to build communal irrigation facilities with improved permanent facilities needed for increasing productivity and thus, it was necessary for government to extend assistance in the construction or improvement of such systems. Permanent dams and structures were needed to replace temporary ones that required seasonal reconstruction or repair which took away substantial time and delayed farming activities. In many systems, additional canals and structures were needed for expanding the area irrigated. On the other hand, it was also clear to NIA that the assistance should not undermine, but rather strengthen farmer ownership of the system and farmer investment of time and labor to develop commitment to system maintenance. NIA subsumed these objectives under a search for developing a process for maximizing farmer participation in all aspects in the planning and construction of the irrigation systems.

In the search for the process, NIA wanted answers to the following questions:

(1) How can farmer participation be maximized in all phases of the development of an irrigation system?

(2) Does farmer participation result in more viable irrigation associations with greater capability for operating and maintaining the system? If so, how can the processes be developed for broad applicability throughout NIA?

To find the answers, in 1976, NIA established two pilot projects in Laur and Nueva Ecija. One planned to serve 600 hectares and the other 1,600 hectares. Both diverted water from a large river by means of temporary earth and brush dams which were washed away by every heavy flood. Both had existing canals that irrigated about half the area being planned for irrigation. Maximizing participation of farmers in these systems meant developing participation at the grass roots level in the critical activities of: (1) decision-making within the irrigation association, (2) planning improvements and expansion of the irrigation systems, (3) securing water rights and rights of way for canals and other facilities, (4) construction of irrigation facilities, and (5) control of construction costs. Construction would start only after the farmers had been organized to participate in these activities and fully understood and agreed on the components of the projects to be undertaken.

Farmer Participation: Community Organizers Programs

To achieve this, NIA fielded six community organizers (COs) under a program coordinator. The COs were carefully selected on the basis of ability to communicate with the farmers and their long-term commitment to the project. All were college graduates in the social sciences with some experience in working with the rural and urban poor. Before being fielded, the COs were trained in the basic organizing procedures of integration with the farmer community, groundworking, mobilization of farmers, action and reflection. They were instructed on the policies and procedures of NIA and exposed to an operating irrigation system to understand its components, functions, and problems. Then they went out and lived in the farm communities in the two pilot project areas. For integrating with the farmers, they joined planting and harvesting activities, social functions, church festivities, and conducted house-to-house visits. After gaining acceptance of the community, they encouraged the farmers to discuss their irrigation system and its problems, the need for expanding the service areas, maintaining the system, and strengthening their associations. Throughout their work, the COs acted as catalyzers in analyzing, persuading, and arguing, but they never took the decision-making away from the farmers, nor performed the tasks which the farmers themselves should undertake. At the same time, they informed the farmers about the assistance that they might obtain from NIA. In this manner, the kind and extent of assistance was defined, and negotiations were subsequently conducted by the farmers with NIA.

After completing negotiations with NIA, the farmers in the smaller system, with guidance from the COs, organized committees for surveys, right-of-way acquisition, revision of by-laws, registration of the association with the Securities and Exchange Commission (SEC), labor mobilization, materials checking, water permit acquisition, and control of construction costs. For developing grass-roots participation, the COs divided the service areas into sectors and farmers selected their sector leaders. Sectoral meetings were held for revising and ratifying the association by-laws. The farmers registered their association with the SEC and secured their water permit from the National Water Resources Council. The same process was used in the larger system, although it took much longer due to the larger area and the conflicts within the existing association that repeatedly frustrated full grass-roots participation.

In the participatory process developed in the smaller system, farmers prepared a map of the proposed irrigation service area, indicated where they would like the canals to pass, discussed these routes with the engineers, and accompanied them during the surveys for the planning and design of the system. Preparatory to construction, discussions were held on the type of diversion structure, construction schedule, extent of farmer participation in the construction work, and the estimated costs that farmers would repay. A major expense item was the diversion weir on the river. Farmers wanted a permanent concrete structure at the site where they built a temporary weir with much effort and expense every planting season, but the funds available for the project were not adequate for such a structure. Thus, an agreement was reached on a semi-permanent gabion weir. After all components of the project were agreed upon, the irrigation association and NIA agreed in a contract that NIA would improve and expand the irrigation system and the association would operate and maintain the system and repay the cost of construction in accordance with NIA policy. In the case of the larger irrigation system, negotiations for construction were deferred until after the internal conflicts were resolved, extensive grass roots participation developed, and the association adequately strengthened.

During the construction, the association mobilized labor of its members to help in the acquisition of right-of-way for canals and stake-out of canal lines. A substantial part of the labor in the diversion weir and canals was undertaken by the association members, without payment from NIA, as the counterpart contribution of the association. Various committees of the association checked the use of construction equipment, the consumption of gasoline and fuel, the quality and quantity of materials being furnished for the project, and the procurement costs of these materials. As these procedures had never been tried in NIA before, there were initial difficulties. At first, the technical staff disliked these procedures which seemed to be encroachments on their authority, but the supportive attitude of the project engineer removed the difficulties. Meetings were held with the association and the new procedures were adopted.

Progress in the larger system was much slower. At the start of the organizing process, strong pressure from a local politician repeatedly frustrated grass-roots participation. The situation deteriorated when confrontational tactics were employed in the organizing process to counteract the political pressure. NIA deferred construction of the project until after the internal conflicts in the association were resolved and sufficient grass- roots participation had developed.

The pilot projects were regarded by NIA as learning laboratories from which it could develop processes for maximizing participation of farmers and learn about its problems and effects. The work attracted the interest of a number of institutions associated with NIA in other activities. These were the Institute of Philippine Culture (IPC), the Asian Institute of Management (AIM) and International Rice Research Institute (IRRI). The IPC had been doing social research on indigenous Philippine irrigation systems for NIA and had a continuing interest in improving irrigation associations. AIM was developing a course on the management of rural development and saw the importance of farmer participation. For a number of years IRRI had been working with NIA on irrigation research and was strongly interested in developing irrigation associations for better water management. These three institutions had common interests in developing farmer participation. Together with NIA and the Ford Foundation they agreed to form a working group to assist NIA in the improvement of the communal irrigation program based on the farmer participation approach. By early 1979, a Communal Irrigation Committee was officially formed by the NIA Administrator with Benjamin U. Bagadion as Chairman. Invited to join as additional members were the FSDC, the University of the Philippines at Los Banos, the Economic Development Foundation, and later, the Central Luzon State University. The CIC was able to draw from members' expertise in engineering, agriculture, sociology, economics, anthropology, institutional management, and training. The CIC met at least once a month to review progress, address problems and agree on activities.

To assist the learning process, the CIC requested the Institute of Philippine Culture to document the processes and activities undertaken in the pilot projects. The monthly reports of the documentation were discussed during the CIC meetings along with issues and problems that had long-term implications for the agency and the program, such as lessons that would be useful for program expansion, and training needs of COs, engineers, technical staff, and irrigation association leaders and members.

Farmer Participation: Lessons Learned in the Communal Schemes

The COs stayed for about three years in these pilot projects; at least ten months of which were for integrating with the communities, organizing the associations and guiding the activities of the associations prior to construction. Near the end of the three-year period, an assessment of the activities showed that while there had been many problems and still much room for improvement, the results were generally satisfactory. The basic processes for inducing farmer participation in planning and construction of communal irrigation projects had been developed and many lessons had been learned, among which were:

(1) Sufficient lead time should be given the COs for organizing and mobilizing the farmers before construction. In the 600-hectare project, this required ten months. During that time, they participated in planning and pre-construction activities.

(2) Engineers and other technical staff should be trained to develop flexibility in their attitude towards farmers, and to gain a basic understanding of the processes being used by the COs.

(3) Engineers and COs should work together closely and integrate the technical and organizing activities into one process.

(4) Agency policies and procedures that obstruct farmer participation should be discarded or amended.

(5) Farmers extensively participate in planning and construction activities that they find beneficial.

(6) Farmer participation, when properly harnessed, has potential for improving the planning of a system. It enables farmers to provide information on canal locations that are more in accordance with their needs. Furthermore, it reduces government expenditure in the construction by the labor and materials contributed by the farmers as their counterpart in the project.

While generally the Laur pilot projects attained the objective of learning about the participation process, the CIC did not consider it as the basis for expansion of NIA's program, for there were some weaknesses to be remedied. In April, 1979, to further improve the process, the CIC established another pair of pilot projects in the province of Camarines Sur where the general conditions and characteristics of the farmers were different from those of Nueva Ecija. The two projects followed the strategy and method of organizing used in the first pilots, but improved and supplemented along the following lines:

(1) To avoid the problems experienced, the confrontational elements in the processes were modified.

(2) A project selection method was introduced wherein technical and institutional information in several proposed projects were analyzed and discussed in a workshop attended by engineers, community organizers, and members of the CIC.

(3) A flow chart was developed synchronizing the technical and institutional activities preparatory to construction within a period of nine months for the guidance of COs and engineers on the effective integration of their activities.

(4) Problem areas that needed improvement to promote farmer participation were identified and resolved. Among these were procurement and contracting procedures, preparation of paddy elevation maps of farmer fields, and funding procedures that would facilitate preparatory technical and institutional work.

(5) Initial guidelines on financial management and water management for the associations were developed in consultation with the farmers.

By the end of 1979, the processes developed were sufficiently satisfactory for expanding the program to the twelve regions of the country.

Change

In December, 1979, NIA top management called the NIA Regional Directors to a conference for expanding the program. The Regional Directors and the CIC agreed to start a pilot project in each of the twelve regions. Each pilot project would be a *learning laboratory* in which regional staff capability for using the participatory approach would be developed. Thirty community organizers were selected through a rigid screening process and then trained for three weeks in the basics of irrigation and the new NIA organizing process. With assistance from IPC, the project selection method used in the Camarines Sur pilot projects was developed into a socio-technical profile writing process for communal irrigation projects. NIA personnel were trained in the data gathering and writing of socio-technical profiles for candidate regional pilot projects. With assistance from AIM, and using the lessons and experiences in the pilot projects, engineers were given orientation on planning and construction policies and procedures that promote participation of farmers.

Workshops were held for engineers and community organizers to discuss and analyze candidate project profiles. Members of the CIC, engineers, and community organizers with experience in Nueva Ecija and Camarines Sur pilot projects served as resource persons guiding the workshop participants. The outputs of the workshops were work schedules for each pilot project integrating together the engineering and the institutional activities. The first group of community organizers in the program became the regional supervisors and trainers of the COs in the regional pilot projects. In each region, additional socio-technical profile writers were trained and a program was started for preparing the socio-technical profiles of future projects.

As the construction of the regional pilot projects proceeded, the CIC began thinking about the post-construction assistance that NIA should give to the communals. A systems management working group composed of CIC members and NIA staff was formed to work on the problem. The group developed a training scheme for associations to develop their own systems management plan. The process consisted of a series of short workshops instead of the usual ready-made plans. The scheme consisted of separate modules to train associations on the development of a: cropping calendar, normal water distribution plan and crisis water management plan, conflict management plan, maintenance plan, farm level facilities plan, and the duties and responsibilities of members, officers, and systems personnel.

In preparation for program expansion, NIA top management, with assistance from AIM, trained the NIA provincial irrigation engineers on the new perspectives and processes needed for implementing the participatory approach. Other staff at various levels were also trained on how to properly support the program. The IRRI representative to the CIC helped improve training methods and materials for irrigation systems management and crop production needed for expanding the program. He also helped train engineers in farm level facilities design and paddy elevation mapping for improving water management. The Ford Foundation continued to provide financial assistance and encouragement to the program, providing more than $1 million over a ten-year period. Its program officer provided advice and assistance to NIA and other members of the CIC.

In 1981, a year after the start of the regional pilot projects, an additional province was added to the program in each region to bring 24 out of 68 NIA provincial offices into the learning process. In mid-1981, a World Bank Team appraised a proposed project for assisting communal irrigation development in the Philippines. The team analyzed and assessed the effects of the participatory approach program and recommended its adoption in all communal irrigation projects of NIA. In 1982, the program was expanded to all 68 provincial offices and by 1983 had become the standard procedure in all communal irrigation projects of the agency.

Farmer Participation: The National Systems

Change also moved into the national systems. The operation and maintenance of these systems has long been a problem of government. The intent, as laid down in policies, is to fund the costs of O&M from irrigation fees of the water users. However, annual fee collections have always been much lower than annual expenses. From 1980 to 1985, NIA expenses for O&M of national systems amounted to $688 million, while collections totalled $541 million against billings of $906 million for that period. (These expenses are only up to the level of the Irrigation Superintendents of the system.) The deficit had to be covered by NIA through subsidies from other funding sources. NIA top management realized that the subsidies could not continue indefinitely. Previous experience showed that raising the level of irrigation fees was not a solution to the problem. Alternative solutions were required.

A scheme that had always attracted NIA was to hand over financially unviable national systems to irrigation associations. These generally were the small systems, where even one hundred percent collection would still be less than NIA's O&M expenses. For in a small national system, there was a minimum number of staff members needed for NIA's operation, whose salaries and fringe benefits, together with other system maintenance expenses, exceeded what possibly could be obtained from a hundred percent collection of irrigation fees (at prevailing rates). A small, national system could be operated just as well, if not better, by a well-organized and well-trained farmer irrigation association at a much lesser cost. Another scheme applicable to a larger national system was to hand over to irrigation associations the O&M of discrete areas supplied from different lateral canals. This left the O&M of the source of supply and the main canal to the NIA, thus reducing NIA's expenses. These schemes would depend on a tested strategy and process for organizing strong and viable irrigation associations. The processes developed in the communals appeared to be the answer to the problems of the national systems.

In December, 1980, community organizers were fielded in the Buhi-Lalo Irrigation Project in Camarines Sur, a national system to be improved and expanded from 1,000 hectares to 3,000 hectares. The COs applied the participatory approach developed in the communals with some modifications. The COs lived in the villages and integrated with the farming communities, then commenced groundworking and mobilization. In the existing part of the systems which was to be rehabilitated and improved, farmers reviewed the layout of terminal facilities proposed by engineers, walked through the farm ditch locations with designers, discussed changes in canal locations, and constructed the canals suitable for manual labor. The COs assisted the farmers to organize small groups by turnout service areas. Each small group became a construction unit with a group leader. Each unit was awarded a construction contract for canals and canal structures depending on its capability to develop cooperation and decision-making skills, and by testing identified leaders. As construction progressed, the COs assisted the farmers in organizing the small groups into three zone associations, covering about 1,000 hectares. The associations negotiated with NIA the manner of sharing O&M responsibilities in the system and the corresponding sharing of irrigation fees collected from the farmers.

In mid-1982, the three zone associations initiated negotiations and then entered into an agreement with NIA for operating and maintaining their respective zones and collecting irrigation fees from the farmers. Pending completion of the additional area of 2,000 hectares, NIA had responsibility for the diversion weir and the first 1.5 km. of main canal. A mutually beneficial system of sharing irrigation fees was worked out. The final formula was to give the association 35 percent of the collections up to 50 percent of the seasonal billings, and 65 percent of the collections over 50 percent of the seasonal billings. Post-construction assistance was extended by NIA to the association through systems management training and financial management training. As in the communals, the process documentation was used for improving the work to be used in other national systems.

With the successful processes of the Buhi-Lalo project, NIA top management expanded the program. National irrigation systems under rehabilitation and improvement with World Bank funding assistance, and problematic surface pumping systems were selected. Additional community organizers were recruited and trained, and relevant NIA supervisory staff were trained. Irrigation superintendents and project engineers in charge of the systems covered by the program were trained on how to respond positively to farmers participation. In mid-1986, the program covered about 35,000 hectares in 37 national irrigation systems. Nine of these systems had been fully turned over to farmer irrigation associations. The rest were under joint management; NIA was responsible for the diversion weir and the upper part of the main canal, and the irrigation associations were responsible for the remaining portion of the system and irrigation fee collection. Where the system has been fully turned over to the farmers, the irrigation association pays for the cost of construction in accordance with the policy for communal systems. In the case of joint management, the association turns over all collections of irrigation fees to NIA and the collections are shared in accordance with an agreement similar to that established for the Buhi-Lalo Irrigation Project. Modifications are made on percentage sharing to suit the situation and condition of each irrigation system. As in the communal systems, all irrigation associations were given training in irrigation systems management and financial management.

Future Direction

So far, only a part of NIA has been transformed for maximizing farmer participation in irrigation development; but it is the major part. It includes all provincial irrigation offices, the twelve regional offices, the small and medium-sized national systems, and the central office components engaged in communal irrigation development and in operation, maintenance, rehabilitation and improvement of national irrigation systems. The change still has to move into the large national projects and systems. This will depend on NIA top management decisions. A program for expanding what is now going on in many small and medium-sized national systems is needed. The program will require a combination of system rehabilitation and participation of organized farmers in the planning and implementation of the rehabilitation. To effect joint system management with NIA, parts of the system will be turned over to irrigation associations.

In the sectors which have undergone the change, the feeling prevails that this change is necessary. Experiences have indicated the following benefits in the communals: (1) stronger and more responsible irrigation associations, (2) increased counterpart contributions from farmers during construction, (3) ready acceptance by farmers of the completed physical facilities and their financial obligations, (4) better maintenance of canals, and (5) more amortization collections.

In the nationals, the following benefits were realized: (1) improved maintenance of canals and better water distribution, (2) more irrigation fee collections, (3) reduction of NIA O&M expenses such that fees collected exceed expenses, (4) no removal of farm ditches by farmers after construction, and (5) ready acceptance by farmers of their share of responsibilities for O&M.

In regard to NIA staff, the irrigation engineers and technical and institutional personnel who were exposed to the program were able to broaden understanding of their work beyond limited spheres. They realized the need for understanding the roles of other disciplines in irrigation development and the need for developing skills for interdisciplinary activities. The program activities, especially the workshops, promoted wide expression and identification of problems. The learning process approach encouraged fruitful discussion, rather than concealment or dilution of problems and enabled treatment of root causes rather than apparent symptoms.

There are, however, remaining items for change. One is the reorientation of a significant sector of NIA that is undertaking the planning and construction of the large foreign assisted projects. This should not be difficult as the methodologies are now available for handling this, as well as the implementation of farmers participation at an appropriate level in these projects. Another is the problem of displacing NIA O&M personnel when a national system, or a part thereof, is handed over to a farmer irrigation association. Some solutions for this have been worked out, but improved ones are needed. Such displaced personnel have priority for employment in NIA construction projects and, if separated from NIA, are given separation pay as set by the NIA Board of Directors. The third is the development of operational linkages and relationships between irrigation associations and NIA in large irrigation systems where NIA would be operating the main canal and diversion works, while many irrigation associations in various laterals operate the rest of the system. This still has to be worked out and plans for action research are underway.

It is difficult to predict how long it will take to complete the work on the existing large national systems, but if adequate financial resources are made available, it would be possible to cover substantial parts of most large national systems within the next ten years. One may expect that by the end of this century, joint management between NIA and irrigation associations will be the prevailing situation in most national systems, while many more communal irrigation associations will have attained self-reliance and viability through farmer participation and NIA assistance.

9

Irrigation Management by Farmers: The Indian Experience

K. K. Singh

Introduction

The severe drought of the last three years, which brought untold miseries to the rural poor of the Indian Subcontinent and necessitated huge expenditures on employment and rehabilitation programs, has emphasized once again the necessity for water planning and water conservation as part of a comprehensive strategy for national development. Never before was there a better appreciation of the environmental factors that could affect the availability of water, nor how their neglect threatens human welfare today.

India is richly endowed with water resources. Out of the 400 million hectare metres (mhm) of annual precipitation 70 mhm is immediately lost through evaporation. Some 180 mhm constitutes run-off, of which 17 mhm is stored in reservoirs in major and medium irrigation projects (Vohra 1985 : 29). The remaining 150 mhm enters the soil and sustains agriculture and irrigation from groundwater sources. Such bountiful rainfall and long months of sunshine are nature's most precious gifts to the Indian subcontinent, where some 75 per cent of the population is still rural and depends upon agriculture and allied occupations for survival. The unpredictability of rainfall and its uneven distribution have been the main cause of famines and acute scarcity from times immemorial (Indian Famine Commission, 1898, 1901). Some part of the country is affected by drought or floods every year. Rains are particularly meager in some years, thus creating famine conditions. India has been visited by such famines once in every 50 years but their frequency has increased in the last century.

The sizeable recurring expenditure on human rehabilitation and the immensity of suffering from famines pushed the British government, from the second half of the 19th century onwards, to recognize the fact that the abundant but unpredictable water resources of the Indian sub-continent could be profitably harnessed to sustain agriculture in years of lean rainfall. The upper Ganga Canal (1854), Agra Canal (1873), Sone Canal (1875), upper Barib Doab Canal (1879) in the north and the K. C. Canal (1870), Godavari delta system (1890) and the Krishna delta system (1898) in the south were among the first important river diversion schemes to be completed by the British (Indian Irrigation Commission, 1972). These canal and anicut systems brought prosperity to regions they served. Large storage reservoirs were built in the 20th century. Attention was given to rehabilitation of old projects, especially tanks, and to the building of new tanks. The use of groundwater was also encouraged. Deep tubewells and lift irrigation schemes received considerable attention following the extensive electrification of the countryside in the Post-Independence period.

Resources and Threats

At the time of Independence, 22.6 mha were under irrigation of which 9.7 mha were major and medium schemes. By the end of 1985 a total potential of 68.9 mha had been created, 30.5 mha from the major and medium projects and 37.4 from minor irrigation projects. However, a gap of 7.5 mha between the potential created and utilized exists at present. The gap between the created and utilized irrigation potential is expected to increase in the future. This is a matter of serious concern given the rising cost of irrigation development.

With a steadily growing population estimated at 748 million in 1984 and expected to reach 950 million by the year 2000 (Planning Commission, 1985), there is an ever increasing demand on land to feed humans and animals. By 2000 A.D., 240 million tons of foodgrains will be required as against around 150 million tons produced at present. There being a physical limit on cultivable land and water that can be harnessed for irrigation, attention must turn to the productivity of land and water both of which hold the key to social security and political stability. Watershed management with full regard to soil and moisture conservation, the preservation of river basins from soil erosion and deforestation, and careful attention to irrigation management have acquired urgency in the present context. India is said to be losing 2 maf (million acre feet) of live storage in large dams due to siltation. This means a loss of 700,000 acres of irrigated area per year (Kanwar, 1988:58). Waterlogging and soil salinity have already become serious problems in several irrigation projects in the Indo-Gangetic plain and the deep black soils of the Deccan Plateau. In some projects almost as much land goes out of cultivation due to waterlogging as is being brought under irrigation.

The Human Factor

Comprehension of the integral relationship between the landscape, forest cover and run-off, on the one hand, and the performance of the irrigation system for the purpose of agriculture on the other, has been slow to come. Even less appreciated is the importance of the user of the land; that is, the people who inhabit river catchments and those who use irrigation water in making a success or failure of whatever the government implements. Consultations with the people, understanding their needs and reactions, involving them in decisions and encouraging the mobilization of collective effort to deal with problems have been woefully neglected. Most government officials are skeptical about the value of involving people in program planning and implementation (Lowdermilk, 1985:1-5). Often, irrigation management is considered solely the government's responsibility. It is up to irrigation officials whether or not they consult farmers or seek their assistance. Due to this, farmers neither take interest in irrigation management nor are they given any serious responsibility. The loser is the Indian nation, for, the lack of user cooperation is responsible for the waste and misuse of water, higher administrative and maintenance costs and lower collections of revenue from irrigation.

Management of Irrigation Systems in India

Traditional Irrigation Systems

Farmer-managed traditional irrigation systems are found all over the world (Uphoff, 1986a: 14-17). In India diversions of water from rivers and streams, surface flows stored in tanks, and lift irrigation from ponds, wells and rivers are among the most commonly found forms of user-managed irrigation systems. Some of these are so old that it is impossible to establish their antiquity.

Traditional irrigation systems share some common characteristics. Farmers make an effort to procure or acquire water, distribute it among themselves according to some accepted norms, keep the system operational by contributing materials and labor, appoint individuals who perform specialized functions, sometimes employ workers to assist in the distribution of water, supervise the system, make decisions to resolve problems and maintain links with members and the outside world. The performance of these functions requires group cooperation, orderly discharge of duties and social control over members. A group or individual is appointed to provide leadership and is replaced from time to time in consultation with all users. Farmers involvement in the performance of most functions is the hallmark of traditional irrigation systems.

In state-owned systems, whether large or small, farmers are not actively involved in any function except the application of water to fields. Procurement and guarding of water is the function of the agency. Water allowance and the share of individuals is fixed by the agency. Operations and maintenance is an agency function. Maintenance of field channels and structures is expected from farmers but they often neglect it. Accordingly, a sense of collective responsibility for maintenance which is the single most striking feature of traditional systems is all but absent in state-owned irrigation projects. In consonance with the above, the need for collective supervision and social control is minimal. The irrigation agency is inevitably drawn into these areas of activity, even though it cannot be effective without client support.

Until a decade ago, there was considerable ignorance about farmer-managed irrigation systems in India. Only after the experiences from the Philippines, Bali, Java, Taiwan and Spain (Coward, 1980; Notoatmodjo, 1985; Maass et al.; 1986) became known did irrigation officials and professionals in India become aware of their significance. In due course, an immense variety of user run irrigation systems have been discovered and some have been studied in detail. A few of the better known systems are examined below.

Tank Irrigation. The tanks of the Deccan Plateau (which still account for one third of the irrigated area in the states of Andhra Pradesh, Karnataka and Tamil Nadu) have existed since ancient times. They were the principal source of irrigation in the southern states prior to the large state-owned irrigation projects. Large tanks commanding over 100 ha are usually controlled by the state for maintenance and decision as regards the date of operation/closure. Water distribution and maintenance are the responsibility of the users. In south India, the revenue authorities assess and collect irrigation fees. Typically, watermen (*Neeruganti/Saudi*) are employed to oversee the distribution of water. In some states they are employed by farmers and function under the instructions of a management committee appointed by the water users. Watermen follow their own judgement as regards the application of water to crops. It is the responsibility of the farmers to maintain the irrigation system and to manage the farm distribution system.

Diversion Schemes. Farmer managed river/stream diversion schemes for irrigation are quite common all over the country. A weir or bund is erected across a permanent stream to divert water to farm fields. Where the size of the stream is large and the irrigated area is substantial, a committee chosen by all water users looks after matters of common interest. The committee and the appointed officers perform the many functions listed earlier. Enduring self-management is found where water can cover one or two crop seasons. The *Kuhls Phads, Ahars-Paynes* and *Paniyaras* are names used for such schemes in different parts of the country. An interesting example is the Chhatis Mauza Irrigation System (Pradhan, 1984) which irrigates 3000 ha and has a three tier organization for management. The central committee consists of 16 members who look after water allocation, water distribution and maintenance of the main canal, assisted by committees representing five to six villages for overseeing village level problems, and, lastly, village canal committees consisting of farmers who are directly involved in maintenance, water distribution and group cooperation. Since the area under irrigation is extensive, the main canal being 13 kms long, management requires a good system of communication. A strict watch is kept on users to see that individual claim their rightful share of water. Labor for maintenance and routine repairs are contributed by individuals in proportion to their share of water. There is presumably no discrimination based on wealth as rules are applied irrespective of the user's socio-economic status. Another example is the *Ahar Pynes* Irrigation System found in South Bihar and neighboring areas (Sengupta; 1984). *Ahars* tanks are connected with *Pynes* (diversion channels) taking off from small rivulets. Each *Ahar* may serve several small catchments from different openings, managed by the beneficiaries. The social organization for irrigation is organized by *Ahars* and also by catchments. Catchment groups cooperate with one another for mobilizing collective maintenance for the entire *Ahar*. Such days become a festive occasions. The opportunity for enjoyment brings many volunteers for work. The system is quite old and self regulating. Though these systems can greatly benefit from rehabilitation or modernization, one reason for their continued existence is that external intervention has not become absolutely necessary for their performance. Should the government come in, it is likely that the voluntary character of the *Ahar-Pyne* system will cease to exist.

The Phad System. The *Phad* System is between 400 to 600 years old. It is found in Nasik and Dhule districts of Maharashtra where perennial river flows exist. Bunds or *bandharas* have been built to divert water to distant villages through canals. Some canals are 10 kms away from the river, and pass through deep cuts to irrigate between 100 to 150 ha each. *Phads* are a community-managed irrigation system. Typically, a *Phad* is divided into four parts and most irrigators own some land in each sub-division. A management committee is appointed for a term of two to five years in a general assembly of all irrigators and management of the irrigation system is its responsibility. The management committee employs *Patkaris* who patrol the main canal, *Havaldars* who supervise the irrigation system under the guidance of the committee and *Jaglias/ Salonias*, who function as watermen/watchmen. Each *Phad* grows only one crop and the decision to grow crops is taken at the beginning of the irrigation season in May. In years of water scarcity, the water users as a whole may decide to exclude a sub-division from irrigation so that the remaining area may receive adequate irrigation. The *Phad* system is a remarkable organization of human effort, self-sufficiency and keen awareness of external social/economic forces. Maintenance is the most important group activity which is carried out once and, if necessary, twice a year by all water users. Fines are rarely imposed as social compliance is good. Social pressure is used to discipline individuals and to mobilize group effort. Reduced availability of water is placing *Phads* under stress. Better water harnessing and rehabilitation is now required to keep these systems from passing into disuse.

Kuhls of Himachal Pradesh, Jammu and Kashmir. Some three to four thousand *Kuhls* (canals) are said to be privately owned. Like the tanks of south India they are in different stages of disrepair. The traditional arrangement for management has weakened due to the replacement of influential people of the past by representative institutional leadership in rural areas.

Kuhls or canals were constructed to draw water from streams that drain valleys and to carry it to far flung villages, 10 to 15 kms away. Water is drawn from a stream by simple diversion structures. The area irrigated varies from 100 to 150 ha. The rights to water of the villages/hamlets served by a *Kuhl* have been fixed by tradition and recorded in official documents. The system of distribution has been fixed taking into account the needs of the paddy crop which requires heavy irrigation during transplantation. Water needs are supplemented by rainfall, but maintenance is required all through the cropping season. Water supply is staggered between users so as to meet the transplanting requirements of all users.

Water is divided from trunk channels to be roughly proportional to the area irrigated. At the point where water reaches the fields, groups of farmers share it according to need. The *Kohli* is responsible for distributing water to villages. He performs two major functions: first, mobilizing labor, material and supervising repairs in the divergence structures, the main canal and control structure and second, supervising the distribution of water along the main canal. The *Kohli* has little involvement in the distribution of water within the villages as this is the responsibility of the water users. Maintenance also rests with the group of users who receive water from a common source. There are various conventions regarding farmers from different villages contributing to maintenance on trunk channels.

Farmers' Self-Management of Rehabilitated-Old and New Irrigation Works

Most traditional flow irrigation systems are gradually going out of use. Whether it is the tanks in South India, *Kuhls* in the Western Himalayas, the *Paniyaras* in the Shivalik and Garwal hills, *Phads* in Maharashtra or the *Ahars-Pynes* in the Chhotanagpur plateau, siltation, distortions in the conveyance system and excessive transmission losses have brought about a marked reduction in the area irrigated. The decline in social solidarity due to changes in the rural power structure has adversely affected management capability. Government intervention is usually required for rehabilitation. A new basis for self-management that can draw upon the emerging social values and forces in the national political economy is needed. In the mean time, the poor sections of society lose out in competition with the socially powerful and the well connected for whatever irrigation water that is still available. At stake is the welfare of many who are getting progressively less out of the irrigation systems which served them well in the past. How to intervene so as to help improve the situation without disturbing the rural communities' capacity for self-management is the most important question in the rehabilitation of traditional irrigation systems.

Major efforts at rehabilitation of old tanks and construction of new tanks are being made in Karnataka and Tamil Nadu. The intention is to involve farmers from the very beginning so that they take interest in their management.

Farmers' Participation in Lift Irrigation Schemes

The ultimate irrigation potential for minor irrigation is estimated at 55 mha out of which groundwater potential is 40 mha and surface schemes 15 mha. Towards the end of the sixth plan period, minor irrigation contributed more to the total irrigation potential at 37.4 mha (55 per cent) as against 30.5 mha (45 per cent) from major irrigation. Much of the groundwater potential (87 percent) is attributable to wells and shallow tubewells which are largely under private ownership. The contribution of the state to the ground water development is limited to deep tubewells and similar schemes. Ground water and small surface schemes are much cheaper to exploit and their ill-effects are considerably less compared to major irrigation. Moreover they provide a wide scope for beneficiary participation. Salunke's *Pani Panchayats* in Maharashtra have shown how much can be gained from the development of micro watersheds to help the poor who inhabit wastelands. In the following section farmers participation in some lift irrigation schemes that have been documented, are discussed.

People's Participation: Public Tubewells

The World Bank funded public tubewells hold the beneficiaries responsible for water distribution and field system maintenance in fulfillment of a condition imposed by the Bank. In Uttar Pradesh (UP), every public tubewell has a Tubewell Area Management Committee (TAMC) which supervises water distribution and looks into disputes or demands referred to it by farmers. The TAMC has an elected Chairman, a secretary and three members. The day-to-day problem of allocating turns is looked after by the Area Day Committee (ADC) of which there are seven corresponding to each day of the week. The farmers under the jurisdiction of an ADC receive water on the appointed day (Singh, 1987 a). An intensive study of five public tubewells in Uttar Pradesh shows that the TAMCs rarely meet, farmers do not maintain the irrigation system, and complaints of farmers are only occasionally attended to by the irrigation staff. In sum, despite the attempt made to involve farmers the response has been poor. Public tubewells function well wherever the government officers are well motivated, farmers control the tubewell technology, or the local leaders take a keen personal interest. By the end of 1989, UP will have 3000 public tubewells with as many TAMCs commanding around 7 *lakh* (100,000) acres. But The performance of these tubewells is likely to deteriorate progressively, without deriving any benefit from the attempted involvement of users in management. The existing trend to increase in hours required to irrigate a given unit of land is an indicator of deterioration in systems management.

Tribal Lift Irrigation Societies

The Sadguru Water and Development Foundation works in the tribal belt of Panchmahal in Gujarat. Its 35 lift irrigation societies command more than 8,700 acres. The work was initiated in 1979 and by 1987, 35 lift irrigation cooperatives were established, most of which are running successfully (Jagawat, 1986). Each scheme irrigates between 60 to 300 acres. Electrically-operated pumps lift water 40 to 150 ft. from the source to the farmer's field. A management committee consisting of the water users is elected every three years. The committee has a honorary chairman and a paid secretary. Meetings are held regularly. The management committee appoints a pump operator, a water distributor and sometimes a watchman whose salaries are met out of the contributions of all water users. Although the farmers are small land holders, (irrigated holdings varying from 1 to 2 acres per capita), they promptly pay irrigation charges and cooperate in all collective efforts. The Foundation supervises the functioning of each society by monitoring its performance through the secretary. Major breakdowns are attended to immediately so that losses are minimized to the extent possible.

Decisions regarding days of operation, sequence of water supply, adequacy of irrigation, collective maintenance, collection of water charges and the enforcement of discipline are taken by the management committee. The two villages from which income data were collected showed that per acre income was around *rupees* 1550.

Public Tubewells and Tribal Lift Irrigation Societies: A Comparison

Public tubewells and the tribal lift irrigation societies stand in contrast. The latter were more costly to operate, the farmers, on the whole, were poorer and less adept in agriculture. They were unfamiliar with the culture of the civic society. Yet, they could make a success of a scheme which could not have been run without full cooperation between members. The public tubewells were installed by the government. Due to subsidy, water rates are not terribly high and the operation of the system did not depend on the managerial skills of the farmers. And yet, what little group effort was required from the farmers to manage and maintain the distribution system did not come forth. The tubewell administration invited farmers participation but did not make a determined effort to sustain it. In case of the water cooperatives, however, participation was carefully nurtured and firmly established.

Lift Irrigation Societies for Cash Crops

Lift irrigation schemes owned and operated by the government do not work well. In the same area, a farmer-run group lift irrigation scheme will most likely succeed. Successful lift irrigation societies generally produce cash crops such as sugarcane, vegetable or fruits. These crops give economic viability to the society. Many such societies, cooperative and others, have been set up with the assistance of financial institutions and refinanced by the state. Technical experts are invited by farmers or made available through the offices of financial institutions. Most societies are able to clear off loans and in some cases sizeable reserves have been built. The *Panchanganga Sahakari Pani Purvatha Mandal*, Kolhapur, which has 650 members covering an area of 607 acres, paid off all loans within ten years and has accumulated a reserve of *rupees 8 lakhs* (NABARD, 1982). Similarly, the lift irrigation schemes at Ujjain Dam Backwaters is another example of successful operations with sugarcane (Datye & Patil, 1987:143). The Shivtakrar Community Lift Irrigation Scheme, another successful venture, owes its management capability to the support of a voluntary agency and its economic viability to onions, as a cash crop, and dairy farming (Datye & Patil, 1987:176).

Lift Irrigation Societies for the Poor

Voluntary organizations, social workers, internationally-funded welfare organizations and some individuals deeply moved by the suffering of the poor have on their own or in association with others, formed societies of water users. The intention is to enable the poor to grow several low water consumptive food crops to meet their needs. State technical and financial assistance is provided for installing the lift irrigation system. Beneficiaries are made responsible for the subsequent management of the system. Follow-up assistance is provided until the group becomes self-sufficient. Records of area irrigated, expenses incurred, water charges, recoveries, and loan amount paid are carefully kept.

The most notable achievement has been V.B. Salunke's *Pani Panchayats* established in a chronic drought area of Maharashtra. Salunke's scheme attempts to harness the water resources of a sub-watershed for the benefit of the poor. Starting with preliminary exploration for establishing the technical feasibility of his idea in 1974, the first *Pani Panchayat* was formed in 1979. By 1985 there were 47 lift irrigation schemes irrigating 1275 ha belonging to 1641 farmers. Water captured in percolation tanks is pumped into fields from wells. Each family is entitled to water sufficient to irrigate 1 ha of land. Only low water consumptive crops are grown. The poor have greatly benefitted as attested by the popularity of the *Pani Panchayats* (Deshpande, 1986).

A different approach to meeting the needs of the poor is illustrated by tubewells owned and operated by the small and marginal farmers in the water rich plains of east Uttar Pradesh and Bihar (Pant, 1986). One such effort in Uttar Pradesh is supported by the Indo-Norwegian Agricultural Development Project and the People's Action for Development. It is implemented at the district level by a Project Implementation Committee which is registered under the Societies Act. Its members are district level officers, representatives of the Indo-Norwegian Project and workers of the People's Action for Development. The work of organizing farmers is carried out by the staff of the Indo-Norwegian Project.

The management of each tubewell, which on the average irrigates about 10 acres or so, is the joint responsibility of the farmers and their leaders. Water is supplied to members and non-members, about 30-40 under each well. Each tubewell costs about *rupees* 15,000. There were 42 tubewells covering 21 villages in 1985. Records of irrigation are kept by the operator of the tubewell who is usually the leader of the group. Most wells were reported to be working well, although there were difficulties regarding adequate supervision and the repayment of loans. Many farmers were defaulters.

Farmers' Participation in Government-Managed Large Irrigation Projects

The extensiveness and diversity of farmer managed irrigation systems whether tanks of peninsular India, the *Kuhls* of the western Himalayas, or the numerous river and stream diversion schemes found all over the country, clearly establish the fact that farmers are capable of both creating and managing irrigation systems. Any doubts in this regard are untenable in spite of the difficulties that have been faced in creating farmers' organizations under government owned large irrigation projects. The ancient grand anicut on the Cauvery river that irrigates 0.24 mha and the Vijayanagar channels constructed between 13th and 16th century in Karnataka are good examples of farmer managed large irrigation systems. The 16 irrigation channels totalling 217 kms in length still irrigate some 29,000 acres in the Vijayanagar system. Although these channels have now been taken over by the government, farmer participation is still in evidence for repairs and the distribution of water at the farm level.

The British were aware of the advantages of supplying measured quantities of water and levying water charges by volume rather than area irrigated which had to be measured every year. Some efforts at volumetric supply on the Ganges Canal were made around 1854 but these trials were given up in a few years on account of the difficulty voiced by farmers with regard to the internal distribution of water among water users and doubts about the amount actually consumed (Indian Irrigation Commission, 1903). Rotational water supply with irrigation time fixed for each farmer was a way to bring about more orderly supply and to cut down waste. When the major irrigation projects were being constructed in the previous century, the main problem was generating full demand for irrigation resources already created rather than conserving water for more extensive coverage of agricultural land.

The urgency with which the problem of efficient water utilization is viewed today did not exist in the distant past. The village social structure gave local leaders considerable influence over the common folk. A system of irrigation administration was developed wherein the government functionary and the influential local leaders could, between themselves, deal with most problems of irrigation at the farm level.

Outlet Committees

In more recent times, attempts at exploring the scope of farmers participation in the management of large irrigation projects with a view to promoting better use of water dates back to 1973-74 when the concept of command area development was introduced. Command area project authorities had been given the responsibility of providing water to each farm field in a bid to raise farm output. Project administrators realized that on-farm development and farm water distribution could be immensely improved with the support of farmers. Some administrators began to experiment with ways to elicit farmers involvement in irrigation development (OFD) and irrigation management (*warabandi*). The outlet command was the unit of cooperation. Typically, farmers were informed, meetings called, views obtained and support sought. Notable success on a limited scale was obtained in some projects in Andhra Pradesh (Sri Rama Sagar Project), Maharashtra (Girna and Mula-Kukadi Projects), Gujarat (Mahi Kadana Project) and Rajasthan (Chambal and Indira Nahar Pariyojana). A common experience was that though initially farmers felt quite involved they lost interest when the OFD works were completed and RWS introduced. This was particularly striking after the irrigation agency officials responsible for program implementation had been withdrawn for assignment to new areas. What is more, indifference towards the farm irrigation system gradually returned. Maintenance was neglected, damaged structures were left unattended and at places willful damage was committed. All in all, there was marked deterioration in the condition of the farm irrigation infrastructure and the water distribution system. The initial gains observed in working with farmers were soon lost. In many projects the expenses incurred on the improvement of the on-farm systems have been wasted due to neglect and disuse. Only where the project authorities were vigilant and gave importance to water users have the systems been preserved.

Experience with Outlet Committees

Outlet-based irrigation associations do not survive unless supported by frequent contacts and follow up from the irrigation agency. In many states where farmers organizations (variously called pipe committees, *Kolaba samitis* or *Pani Panchayats*) were created, they either were still-born or withered away with time. For example, in the Sri Ram Sagar (Pochampad) Project, nearly three thousand outlet based pipe committees covering 1.2 *lakh* acres were functioning at one time and at least half of them were fairly healthy. But with the change in leadership at the project level and the transfer of the field staff these organizations were adversely affected.

When first created, outlet committees neither have social cohesion to bring a collective approach to bear on issues of common interest nor do they have the capability to deal with shortages in water supply. Being the last unit of the hydraulic system, outlet-based farmers organizations find it difficult to tackle inadequate or unpredictable water supplies. The *Pani Panchayat* in Maharashtra, which was organized at the minor level, was better able to deal with the water requirements of the outlets, but it was used as an arm of the government to carry out functions assigned by the irrigation agency. It did not function as an autonomous body of farmers having an agenda of its own.

Experience suggests that the area to be managed by an irrigation association should be as self-contained as possible. The association should be in a position to monitor water supply at the transfer point and influence the irrigation agency to ensure full supply. Only then can it distribute water among the users and establish its credentials as an effective organization. Whether the transfer point should be the minor or a larger hydraulic unit will depend on the confidence with which the irrigation agency can deliver water at a particular point.

Farmers must feel that they gain by joining an irrigation group. One powerful incentive for group cooperation is receiving one's due share of water with an acceptable degree of certainty. Assured supply is the main reason behind the success of the north Indian *warabandi* system. But there it is the government that assures delivery rather than the farmers. And, since the government could not oversee *warabandi* in the entire irrigation command, the *warabandi* system has fallen apart in many locations.

Besides certainty, which is the primary incentive for group cooperation, some other incentives may also attract farmers into forming groups. For example, the prospects of a well-maintained irrigation system, fair assessment of water tax, a just forum for the exchange of information and the settlement of disputes, and an atmosphere of cooperation between water users can all attract participation. Social viability, namely, emergence of the right kind of leadership and decision-making, consultation with irrigation, enforcing discipline with group support and maintaining good working relations with outsiders are important contributors to self-sustenance of a group. Taken together, these give a legitimacy to the farmers organization and establish its credentials as a legitimate forum for water management.

Farmer organizations must also be economically viable. The significance of this point has been completely missed in outlet-based committees. Farmer organizations have to incur expenses on bookkeeping, employment of workers and occasionally on repairs. The area covered by an association should be large enough to yield income adequate for these purposes and some savings for the future. Where water supply is volumetric, the difference between the rate at which water is purchased from the agency and sold to users can yield savings sufficient for meeting common expenses. The water users cooperatives and lift irrigation societies referred to earlier follow this practice. The farmers should feel that by joining a group of irrigators, they gain individually. Otherwise, the motivation for group management will be weak and the irrigation associations will need assistance from the official agency to sustain them.

Water Users Cooperatives

Outlet committees can be made effective provided they are supported by a two or three tier organizational structure, (for example one at the minor level and another at the distributary). Each one of the three will have to function as an integrated unit having responsibilities appropriate to their level, backed by requisite powers. Another alternative is to create a unit which is reasonably self-contained both as a hydraulic and social unit. The water users cooperatives set up in the Ukai Kakrapar canal system in Gujarat are excellent examples. Water is supplied by volume and the geographical area is fairly large to make these cooperatives economically viable. Being located in one village, but sometimes serving several continguous villages, they have the additional advantage of being able to acquire legitimacy as social institutions for performing a function quite central to the needs of people following agriculture as a occupation.

The Mohini Water Users Cooperative is the best studied so far. It covers an area of 487 ha drawing water from four sub-minors and two direct outlets (Shah, 1988:465). The farmers have chosen a board of directors consisting of nine members. Sugarcane is the main crop which gives good returns on investment. The difference between the bulk rate at which water is purchased from the government and the rate chargeable to members on crop/acre basis yields savings which helps pay an annual wage bill of *rupees* 27,000 and results in some savings that go as reserves. The salaries of a manager, a clerk, a supervisor and three canal watchmen are paid by the society. Although it has been said that sugarcane has contributed to Mohini's success (Datye & Patil, 1987:4), it cannot be denied that many farmers who now get water had been deprived earlier. Water is also supplied to non-members. It is reported that farmers in Mohini use 40 percent less water than their neighbors in surrounding villages.

Another water users cooperative, Rayma, also on the Kakrapar system, has been functioning for the last six years. It has a Cultivable Command Area (CCA) of 452 ha. Seventy percent of the area is covered under pulses and cotton and the remaining with other crops, sugarcane being less preferred. For several years Rayma was running at a loss when in 1987-88 it made a profit of *rupees* 2000. During its six years of operation, the irrigated area has increased from 69 ha to 239 ha As in the case of Mohini, Rayma farmers on the average, use 40 percent less water. More water users cooperatives are emerging. Their success indicates that water cooperatives on large irrigation projects can be successful. The slow progress with them demonstrates that the task is not an easy one.

What Farmers Contribute to Irrigation

Difficult though it may be to create and sustain irrigation associations as self-managing social entities, they contribute immensely to the performance of irrigation systems. Some of the gains observed from farmers' participation are reviewed below:

Increase in Net Area Irrigated and the Number of Irrigators Supplied Water

Observations from several projects show that the area irrigated increases after the formation of irrigation committees and the introduction of rotational water supply. In Pochampad the net area irrigated increased by 25 to 30 percent in sample commands (Singh 1982:34). In Mula the area irrigated increased from 7.2 ha (1981-82) to 43.2 ha (1985-86) in *kharif* and from 51.8 ha (1981-82) to 111.5 (1985-86) in *rabi*. Similarly, the number of irrigators increased from 17 to 66 and 110 to 162 in *kharif* and *rabi* respectively between 1981-82 and 1985-86. Similar improvements have been noted in the case of Mohini and Rayma water users cooperative societies. Controlled application of water to fields has resulted in savings of upto 40 percent. The irrigation command under Mohini and Rayma cooperatives is still increasing. In the latter it has increased from 69 ha to 239 ha in a total CCA of 452 ha.

Maintenance

The quality of maintenance improved in the Mula project. With greater cooperation between farmers, the leadership was able to mobilize cooperative effort for maintenance (Singh & Firdausi, 1988). In Pochampad, farmers were willing to contribute for maintenance but collective maintenance was not required since the system had been recently constructed. However, an estimate of farmers contribution in monetary terms to maintenance below the outlet was estimated at *rupees* 830 per year for an average outlet command of 50 acres. This comes to a staggering figure of *rupees 2.7 crores* 10,000,000 for the entire command of 16.2 *lakh* acres when it is fully developed (Singh 1982:44). Moreover, farmers can take some responsibility for maintenance of minors whose upkeep is often neglected due to inadequate funds at the disposal of the irrigation agency. There is enormous scope for better maintenance at lower cost provided farmers decide to take the initiative.

In the Philippines, water distributors (ditch tenders) are not required where farmers have taken up the responsibility for irrigation management. When water distributors retire or are transferred, they are not replaced. This has led to savings in establishment expenditures.

Water Sharing and Conflict Resolution

A distinctive gain in the Pochampad project was the early settlement of disputes (Singh 1982:62-68) which would otherwise have dragged on and created ill will between water users. The same was reported in Mula. The *Pani Panchayat* was instrumental in helping farmers adhere to the irrigation schedule so that disputes about one's rightful share did not arise. In Mohini and Rayma, the chairmen of the water cooperatives are invariably consulted when an issue arises which the concerned parties cannot resolve on their own. Earlier, disputes had to be referred to the irrigation officials, and, pending resolution, they remained unsettled.

Close Supervision by Irrigation Staff Not Required

Mula reports that once the *Pani Panchayat* takes over irrigation management, it is not necessary for irrigation officials to personally supervise water distribution in villages. Leaders collect applications from farmers and return them after they have been sanctioned by the irrigation officers. This has enabled the irrigation staff to visit more villages and to attend to problems where their presence is necessary. In water cooperatives the presence of agency officials is not required. The society manages its affairs with the help of farmers. Official intervention is required only in the matters of water supply, repairs to canals and illegal tapping by upstream farmers.

Collection of Water Tax

Water users cooperatives collect water tax from members. The managing committee is ever vigilant in this respect. In Mula the *Pani Panchayat* was instrumental in persuading farmers who were defaulters to pay off their tax. In Pochampad, since water tax is levied for wet and irrigated dry crops by the revenue authorities, the involvement of pipe committees leadership was not brought to notice.

Increase in Farm Investment and Incomes

Farmer involvement in irrigation has a decided impact on total crop area, cropping pattern and the use of agriculture inputs, particularly seeds, fertilizers and pesticides. In Pochampad, the cropped area increased by 17 per cent after the introduction of *Warabandi* and pipe committees (Singh, 1982:38). There was a reduction in the area under paddy by 33 percent as paddy was discouraged. The increase in income from agriculture was estimated at 65 percent according to government sources. Detailed figures are not available for Mula, however farmers did report larger area under irrigation and improvement in income. The more significant achievement in Mula was the adoption of improved agricultural practices under the influence of the nearby agriculture university. The members of the *Pani Panchayat* used to inform farmers about the latest developments in agriculture and encouraged their adoption. Similar reports have come from the Rayma water users cooperative and the action research project on a distributory on the Mahi Kadana project, both in Gujarat.

The benefits from involving farmers are multi-faceted. Farmers organizations make it necessary for the government agency to consult them and in the process a better relationship for mutual assistance develops. The cost of involving farmers is not high. The single most important costly item is investment on the improvement of the main system to deliver the desired quantity of water at the transfer point (where farmers take over). In most irrigation systems in India sizeable investments had to be made. In the Pochampad project, for instance, a sum of *rupees* 10,000 had to be invested in 1979-80 before the selected minors became capable of delivering the design discharge. Similar investments had to be made in the case of the Mula project and the water users cooperative society on the Ukai Kakarpar system. Such expenses should not be seen as chargeable to the creation of farmers organizations for they relate to the improvement of the irrigation system which should have existed from the time the system became operational.

Other expenses are nominal. In Pochampad where more than 3,000 pipe committees were established, in Mula where two minor level *Pani Panchayats* have been created and for the several cooperative societies in Gujarat, no additional staff was employed and no extra expenses incurred. However, if institutional organizers are employed, as in the Philippines or Sri Lanka, or a separate cell is created, expenses on establishment, travel and the services of the extension staff are bound to increase. All in all, the benefits of farmers organizations far outweigh the investments on their creation and sustenance. The more critical issue is creating conditions which will make it possible for farmers organizations to become viable and to function in the common interest of all its members.

What Prevents Farmers' Participation in Irrigation

The Government of India (GOI) has not failed to recognize the gains from farmers' participation. The sixth five-year plan document observes:

> Farmers should be associated closely in the command area development activities, particularly in land levelling and shaping, construction of field channels and distribution of water equitably. This can be attempted if in each village or under a minor, farmers' association is formed along with representatives of the irrigation and CAD departments. In order to ensure optimum use of land and water, strong and sustained linkages need to be developed amongst the canal management authorities, Command Area Development Authorities and the Farmers (Planning Commission, 1980:157).

The Seventh Plan also gives importance to the subject.

> Farmers' participation in construction activities such as land levelling and shaping, construction of field channels and equitable distribution of water would be encouraged. In order to effect close coordination in the implementation of the CAD programme, representatives of the farmers cooperatives or outlet committees will be fully associated with the programme. (Planning Commission, 1985:82).

The National Water Policy document endorses the involvement of farmers in irrigation and makes a special mention of the role of voluntary agencies in making this possible.

> Efforts should be made to involve farmers progressively in various aspects of management of irrigation systems, particularly in water distribution and collection of water rates. Assistance of voluntary agencies should be enlisted in educating the farmers in efficient water use and water management (GOI, 1987:11).

System Deficiencies

Despite official rhetoric, few new experiments in farmers participation have surfaced during the seventh plan period. Some states, particularly Gujarat and Maharashtra, are making efforts to deliver water volumetrically. This necessarily requires farmers to manage the farm distribution system and perform related functions, including the collection of water tax. In most other states little has so far been done and nothing worthy of note is expected in the balance of the seventh plan period. The Ministry of Water Resources, GOI has offered financial assistance to states for taking up experiments with farmers groups but there have been few takers. The difficulties in achieving a breakthrough seem to result from several sources. The lack of control over the delivery system is one of them. It needs hardly be argued that if irrigation supplies cannot be controlled or supplied with a fair amount of certainty to farmers, the latter will gain little by getting together for the distribution of water or the maintenance of the irrigation system. In fact, uncertain supplies will make cooperation almost impossible.

Rewarding Dependence

The most important reason and perhaps the most difficult to change is the stance of the irrigation agency toward farmers. The agency would like farmers to carry out assigned responsibilities without any concommitant sharing of power and authority. Whether it is farmers associations in Pochampad, Mula or Girna or the *warabandi* system as practiced in the northern states of India, it is inveriably the irrigation authority that sets the conditions under which farmers have to function. In Pochampad, farmer committees were assigned functions. They were never consulted as to what they would like to do and how. In the Mula project, *Pani Panchayats* work at the behest of the irrigation authority. Farmers frankly admit that they are agents of the irrigation agency, and they rather not carry on with the responsibilities any longer. Water users will be willing to take responsibilities for irrigation management only when they stand to gain and when they enjoy the autonomy to influence the irrigation agency according to their requirements. Participation has no meaning if farmers are to work as the last tier of the administrative system.

Disregarding Community-Group Action

The question to ask is why has it not been possible for state irrigation departments to create conditions where farmers can function with a fair degree of autonomy? Why have farmers not been given the power to influence the decisions of the agency in matters of interest to them? And why can they not hold the government agency accountable for providing them services (the agency is employed to perform)? The answer perhaps lies in the fact that legislation for irrigation in India, whether it is the Irrigation Manual Relating to Canals in Bengal (1882), the Northern India Canal and Drainage Act (1873), the Central Province Irrigation Act (1931), the Bombay Irrigation Act (1879), the Hyderabad Irrigation Act (1357 F) and other such Acts, the development of irrigation and its use was conceived as being the prerogative of the government and a gift of the state to the farmers. It was an act of benevolence in the interest of individual farmers. They must, therefore, not only first establish a claim to water and construct water courses at their cost, but never forget that they are the beneficiaries of state munificience. The powers vested in the government by virtue of the irrigation acts look at water users the way kings viewed their subjects. Even when some irrigation acts provide for farmers taking responsibility for the distribution of water as in the Bengal Act and the Central Provinces Act, the primary purpose is not to make farmers self-sufficient or to encourage formation of groups, but to assist the state in collecting revenue from a dispersed population of water users. Both these acts authorize the local influential people to act as tax collectors and to retain a small percentage of the amount collected for themselves, thus creating an intermediary between the state and the water users. The tax collector is in a position to raise taxes provided he paid the state its dues. The irrigation acts recognize the rights of village communities only in the case of irrigation sources that have been in existence and from which the people had been drawing benefits from the past. Assistance to these works, where necessary, was provided without interfering with traditional ways.

Bipolar Negativeness

That the spirit behind the irrigation acts should condition the ethos of the irrigation administration is not unexpected. That this ethos should be reflected in the supervisory styles, work procedures and the behavior of the irrigation staff is only logical. The authority to make decisions backed with administrative and financial powers is vested in the *canal officer* who, in most cases, is of the rank of an executive engineer. But, the officers who interact with farmers over water related issues, who listen to grievances and are approached for solving problems; those most important from the view of the water users, are junior officers belonging to the sub-division and the section. The gulf that separates the sub-divisional officer or the section officer from the divisional officer becomes even wider by the time it reaches the farmers. This gulf or divide is not only one of distance but of information, credibility and distrust.

The survival of farmers organizations in large irrigation projects is totally dependent upon the support available to them from the government. The organization of the irrigation administration, the roles and responsibilities of the officials and their orientation towards water users has to be considered in any long term strategy to sustain management by farmers. At present it is the irrigation agency that runs the irrigation system. Farmers organizations are by and large expected to take over the responsibility for the distribution of water and routine maintenance. The administrative ethos is basically authoritarian. The categories of thinking are bipolar - water can be given and water cannot be given, repairs can be made and repairs cannot be made, adjustment is possible and adjustment is impossible. Negative responses overwhelm the positive. There is no evidence of response based on a spirit of service or accommodation.

Water Users Neglected

In the earlier years of the irrigation administration, it was incumbent upon canal officers, as defined in law, to inspect fields, meet farmers, and redress grievances. Agency authoritarianism was mellowed by a characteristic paternalism which was an expression of the ideal type of the preferred authority figure. Although construction was the main preoccupation then, there was great deal of emphasis on the delivery of water. Irrigation tax was a source of revenue to the government as return on investment (Stone, 1984). In Post-Independence India the importance of operations, that is, the supply of water to farmers has receded in importance as officers spend more time at the table than in the field. Official authoritarianism and its arbitrary expression is no longer counter balanced by benign paternalism or the concern for the delivery of services. Officers are more accessible, even more amiable, but much less responsive to farmers needs than their predecessors. They stand to lose little whether they do the job well or carry on with indifference.

New Laws, New Structures Needed

In designing any administrative arrangement to support farmers organizations in irrigation, it will be necessary to properly interface user groups and their representatives with the irrigation bureaucracy (Chambers, 1984). The whole question of how best self management can be supported with financial, manpower and skill resources so as to improve water utilization, cut maintenance costs, increase irrigation revenue and hike agriculture production will have to be thought through. These cannot be done without restructuring the irrigation administration and changing its sanctions and incentives.

In recent years, changes in the Irrigation Acts of several states such as Maharashtra (1976), Andhra Pradesh (1985), and Gujarat provide for farmers committees whether as cooperatives or elected bodies to make it possible for farmers to take a different role than they have in the past. Similarly, encouragement of volumetric supplies of water, which is endorsed in the irrigation acts of Maharashtra and Gujarat, empower the state to supply water in bulk. But, this legislation will not succeed in improving the performance of irrigation systems unless the department itself is restructured to create new roles, work procedures, the incentives and sanctions which shape bureaucratic behavior (Lusk and Parlin, 1988).

In the ultimate analysis, one must also see whether the elected representatives of the people support farmers participation. Many instances have come to notice where initiative by farmers to gain autonomy and to work on their own, taken at the behest of engineers or administrators, have been frustrated because of the potential threat they pose to the local political leadership. Political support for initiative by farmers is essential. At the present juncture the political climate, the administrative ethos nor farmer willingness favour farmers' self management of irrigation systems. However, a distinct policy change in the direction of people managed systems is discernable in India. Hopes for a better future for all lie in the better use of land and water, preservation of the irrigation system from the watershed down to the farm field and higher agriculture production to meet the needs of a growing population. The people hold the key to the future.

10

Lessons from Small Scale Irrigation Systems in Indonesia

Jeffrey D. Brewer

Irrigation in Indonesia

Indonesia consists of more than 13,000 islands, most of which have mountainous terrain from the volcanic action that gave them birth. In mountains, much of the farmland is concentrated on the floors of small valleys and basins. In addition, the mountains serve as watersheds for small streams and rivers. Small run-of-the-river systems, most serving less than 100 hectares each, irrigate almost all of the irrigated farmland in mountain regions in Indonesia. For the most part, these small systems have been built by local farmers and are still managed and maintained by those same farmers.

This chapter will describe two small Indonesian systems, the village-controlled system found in Bima, and the famous *subaks* of Bali. The object will be to determine the mechanisms that have been developed by Indonesian farmers to allocate and distribute water, to maintain their systems, and to resolve disputes.

Virtually all Indonesian irrigation systems are designed for paddy as the primary crop. Rice, of course, is the staple food in Indonesia. Paddy has some special characteristics. First, it has very high water requirements. Most important for the design of the systems considered here, is that rice can grow in standing or flowing water. Standing or flowing water reduces weed growth and ensures that the plant's water requirements are met. Hence, systems are generally designed to keep water flowing more or less constantly through the fields.

A Bimanese Village Irrigation System

Bima is the name of the eastern half of Sumbawa Island in the Lesser Sundas. Bima is linguistically and culturally distinct from surrounding areas and has over 500,000 inhabitants. All are Muslims and most are dependent upon farming.

In Bima, as in most of Indonesia, villages play a major role in determining social relations. Most Bimanese marry within the village. Village identity is usually a primary identity; there are no groups such as castes that crosscut villages and provide wider groups for identification. Similarly, within villages most people publicly maintain an attitude that all villagers are kin and that all villagers know everything about every other villager.

Villages serve as the local government. Each has a municipal officer who acts as the government head. However, policy decisions and disputes are referred to an amorphous group of men called the village elders.

In Bima, both the court system and higher levels of government support settlement of disputes and control of local affairs by village government. The courts generally refuse to hear cases that can be settled in the village and other authorities usually refer such cases back to the elders.

Although farmland is privately owned, each village has juris-diction over a well-defined area. Within its area, the village government serves to organize public activities and settle disputes. Generally, a farmer who plants land located in the jurisdictional area of another village is subject to that other village's public policies and decisions with regard to farming.

Rasa is one of these Bimanese villages. In 1976, Rasa had a population of 882. It is a farming village, located in a small river valley surrounded by high mountains. Most farm production is from irrigated land on the valley floor, but dry farming on the mountain slopes is also practiced. Most irrigated land is planted with rice three times per year and all of it is planted with rice at least once. The major alternative crop is soy but other crops are planted, most commonly during the dry season.

The irrigation system in Rasa's valley is a river diversion system. Rock weirs divert water into canals that range in length from 5 meters to 2 kilometers. Water is carried by each canal to the top of a block of terraces. After it enters the block, water is led from field to field and eventually drains back into the river. Blocks (commands) range in size from 1.2 hectares to 34 hectares. The thirteen weirs in Rasa's valley irrigate slightly more than 111 hectares. Most terraces are less than 0.1 ha in size and every block includes land owned by more than one farmer. In 1976, the number of farmers per block varied from 6 to 129.

Water Allocation

Water allocation to specific fields is managed by means of division of the irrigated farmland into three classes. Class 1 land is entitled to enough water for rice throughout the year. Class 2 land, mostly located near the ends of canals, is entitled to water throughout the year, but during the dry season may not get enough water for rice. Class 3 land, located at the higher valley edges, usually receives no water during the dry season. About 27 percent of Rasa's land is Class 2 land and only 2 percent is Class 3 land. The remainder is Class 1 land.

Water Distribution

The villagers believe that, for rice, the ideal water control pattern is as follows: First, the fields must be wet for tillage. After plowing or puddling, the terraces are filled and soaked for a few days, then drained for harrowing and transplanting. Water is allowed to flow straight through the terrace for the first few days after transplanting, when the water level is gradually raised, never raising it higher than half the height of the plants, until it reaches 15 centimeters in depth. Ideally, it is kept at 15 centimeters until after heading when it is gradually reduced so that there is no standing water at harvest time. Field water control is handled by regular visits to the field, usually twice a week, but more often during critical periods.

Farmers have limited control of water in the fields. Generally, a farmer can only raise and lower the level of water in each terrace by changing the level of the terrace drain(s). No farmer is allowed to stop the water that flows from his fields into another's fields and no farmer is allowed to shut off the water at the top of the block. A consequence of this system is that, except for those few terraces which can be isolated from the water flows, all farmers in a block must cooperate on crop choice and timing of planting so that they all have approximately the same water needs at any one time.

Coordination of crop choice is not difficult. If others in a block are planting rice, a farmer usually has no choice but to plant rice.

Coordination of timing is accomplished through the village. The whole of the Rasa irrigated area is divided into three irrigation regime areas; one of 70 hectares called *So Rasa*, one of 23 hectares, and one of about 6 hectares (a few small areas are not considered parts of one or the other of the major areas).

For *So Rasa*, the village elders, interested farmers, and specialists in magic meet before each planting season to select a common date for seed preparation. Usually a magic specialist is asked to specify a few particularly auspicious dates on which to begin work, then the group selects the best of these dates based on technical or agronomic reasons. All, or most of the farmers who farm in the regime area begin seed preparation on the selected day; thus ensuring relatively close timing of operations for the regime area.

For the two smaller regime areas, coordination is achieved less formally through discussions among farmers. However, for these two smaller areas, each crop begins about a month and a half after the crop in *So Rasa*. This staggering simplifies time and labor management for farmers with land in two or more regime areas.

Control of water flow from the weir is managed by any farmer in the block as long as he does it in accordance with the group decisions. That is, some farmer will open the weir to reduce water flow into the block at or near harvest. Similarly, once harvest in the block is over, some farmer will repair the weir to begin water flow for tillage of his own fields. Once water flow begins for a season, however, individuals are not allowed to disrupt flow until the very end of the season. There is no formal coordination needed and usually no official looks after these matters.

Maintenance

Like water control, maintenance is often handled by individuals. Maintenance in the fields is always the responsibility of individual farmers. Also, if a farmer notices a difficulty with overall water flow, he usually will investigate and if the needed repairs are minor, such as strengthening a canal embankment at one place or a simple repair of the weir, he will carry it out himself rather than try to get others to do it.

There is a village agricultural officer, called a *punggava*, who is responsible for the coordination of irrigation system maintenance. At least once a year, generally in January, the village *punggava* will call all of the farmers in a particular block together to repair the weir and clean the canal. Each farmer attends or sends someone to work in his place. Attendance is generally good; the *punggava* has the power to fine those who fail to appear. In addition, whenever there has been major damage to the canal or weir, the *punggava* will call all the farmers together to make the repairs. Major damage is most often due to a flood which wipes out a weir.

Many blocks contain farmers from neighboring villages in addition to Rasa farmers. Also, many Rasa farmers own and work land located in the jurisdictional areas of other villages. Any farmer, no matter what his village, must shoulder his portion of the maintenance work. Therefore, when work is needed on a Rasa weir and/or canal, the Rasa *punggava* informs the block farmers from other villages to come at the appointed time. When a significant number of farmers in another village are involved, the Rasa *punggava* will ask the *punggava* in the other village to contact the farmers. Leaders regularly cooperate in this manner.

Resolution of Disputes

The *punggava* is also responsible for settling disputes. He has the authority to levy fines for damages or neglecting work and to order changes in water flows when needed. However, his authority is not paramount and serious disputes will be appealed to the village elders for resolution. The village elders can make decisions that will have the backing of the whole village and, thus can punish transgressors from the village rather severely.

Disputes are uncommon. For example, between September, 1975 and June, 1976 only one irrigation dispute was brought to the Rasa village elders for resolution. The *punggava* settled a few additional disputes without any fanfare.

The frequency of disputes is low because water is not scarce. When crops are in the field, water flows continuously into all of the terraces in all of the Class 1 land and in most of the Class 2 land, even at the driest season. Moreover, when disputes or problems occur, the village government has the means to make a decision and back it up. In most cases, disputants cannot appeal to other authorities.

The Rasa irrigation management system is a smoothly functioning system in which the necessary coordination among farmers is brought about through the use of the village government as the coordinating organization. Rasa farmers neither have nor need an independent water users association.

This system has a cost, however. Having the water flow continuously into all of the Class 1 land means a great deal more water is used than is needed for the crops. Because of the steep mountain walls surrounding the Rasa Valley, Rasa farmers have just about reached the limits of the command area from their weirs. Since there is no alternative use for the water, the water is not sufficiently valuable to make use of it more efficiently. Hence, the decision to take advantage of abundant water to streamline the management system is sensible.

Balinese *Subaks*

Much of the center of Bali, an island just east of Java, is occupied by a mass of tall volcanos. Long ridges descend from the peaks of the volcanos and small rivers flow down the narrow valleys between the ridges.

Bali is unlike the rest of Indonesia in several ways. Its three-and-a-half million people are mostly Hindu. Villages are important units in Bali, but unlike Bima, other social units, such as castes, temple groups, and *subaks*, exist independently of villages.

Tihingan is a village located on one of the ridges in the southeastern part of Bali. The farmers who live in Tihingan all own land in one of the *subaks* in the nearby valleys.

A *subak* consists of the land irrigated from a single weir in the river. Water for the *subak* is diverted by the weir and carried some distance to the head of the *subak*. Then, the main canal is sub-divided in several stages into numerous branches, each one of which irrigates one block of terraces. Over the years, each *subak* has worked out a careful division of water to ensure that all receive equitable shares.

The term *subak* refers not only to the land, but also to the organization of farmers that work that land. All land is privately owned, but every farmer working land in a *subak* is automatically a member. *Subaks* are not small organizations; two *subaks* closest to Tihingan have 222 and 455 members respectively.

Each *subak* has a head and several subdivision chiefs elected by all of the members. These officers supervise all *subak* activities. Every member has one vote in the elections for officers and in general policy decisions. All other obligations are figured on the basis of the amount of land held within the *subak*. The most important obligation of membership is payment of the *water fee*. The funds collected are used to pay *subak* expenses.

Most operations and maintenance activities are carried out by a self-selected group of members called *pekaseh*. About a third of the *subak* members are *pekaseh*. The larger of the two *subaks* near Tihingan has 160 *pekaseh*, while the smaller has 60. *Pekaseh* are responsible for the daily operations and maintenance chores in one portion of the *subak*. Generally they serve in pairs on a general rotation. Each man serves about one day in twelve. At regular intervals, all of the *pekaseh* are gathered together to tackle larger jobs, such as repair of the dam and cleaning of the main canal. In return for this work, the *pekaseh* pay a reduced *water fee*.

Like all other Balinese social groups, each *subak* has its own temples. Generally, a *subak* has a temple to the goddess of fertility in the fields and a temple to the water god at the weir. Regular rituals are carried out at these temples under the direction of the *subak* head.

Water Allocation

Water allocation has two aspects. First, there is the division of the river water among the various *subaks* along the river. This matter is decided by negotiation among the leaders of the *subaks* on the river. Over time, the amount of water to be diverted at each weir has become fixed. In addition, there exist agreed planting schedules that serve to stagger demands on the river water per area.

Water allocation within the *subak* has been decided by the membership. The basic principle is that all land should get equivalent amounts of water.

Water Distribution

The most important aspect of water distribution is the division of water among the subchannels. At each bifurcation, the water flow is divided into fixed shares by means of precisely cut bamboo or wood-notched structures. The proportion of the flow to be passed into each channel is calculated on the basis of the area served by that channel and has the approval of the *subak* membership. One of the duties of the *pekaseh* is to check on the division structures.

In order to even out peak demands on the system, the *subak* regulates the planting schedule, and sometimes even the crop, of each farmer. Each *subak* divides its area into four subsections, each of which adheres to a particular planting schedule. These four schedules are staggered.

Maintenance

Maintenance activities, including channel cleaning and repair of structures, are carried out by the *pekaseh* under the orders of the *subak* officers. It is possible to call all of the members together for a job beyond the capabilities of the *pekaseh*, but rarely is such an action needed.

Conflict Resolution

Disputes are referred to *subak* officers for settlement. Generally, the officers are careful to get the support of the membership for their decisions since all are directly elected by the membership. The officers can fine a delinquent member or, in extreme cases, shut off water to the fields of a delinquent member. Normally, outside agencies will refuse to hear an appeal of a case, hence the *subak* officers have the power to carry out their decisions.

More effort is put into water management in these *subaks* than in Rasa, largely because there is more demand for water in densely populated Bali. The *subaks'* ability to directly regulate planting schedules and even choose crops gives them power to use water in a much more efficient manner than Rasa farmers use water. However, as can be seen, Balinese farmers pay a higher price in the form of cash for payment of the *water fee* and in the form of labor contributions and attendance at meetings and rituals.

Conclusion

These two cases demonstrate that farmers are quite capable of creating their organizations for irrigation management and of managing water in ways that make sense in their environments. Rasa farmers use a great deal of water, but since they have no alternative use for it, it is not wasteful. Also, by using water this way, they manage to irrigate with few disputes and problems and at a very low cost in effort. Water is more valuable to Tihingan farmers because of competition from other farmers. Hence, they invest more effort to make use of water more efficiently.

Both types of organization are remarkably effective. In both cases, part of the effectiveness is due to the fact that farmers have no way to appeal decisions of the organization to outside agencies. At the same time, however, all farmers have a say in the organization's decisions. The presence or absence of these two critical features will strongly affect the performance of any water users association.

Finally, both types of organization are old and in both cases, the systems have had many years to be developed. Rasa's system and village organization are at least 150 years old. The *subaks* near Tihingan are probably older. Not only do the farmers thoroughly understand both their systems and organizations, but also surrounding them are many other examples of similar systems and organizations. The whole environment is, thus, supportive of these organizations. Time for development and a supportive environment are important keys to successful organization for irrigation management.

11

Local Participation in Water Management: The Peruvian Case

Barbara D. Lynch

Introduction

The tradition of participation in irrigation development in the mountainous Peruvian Sierra is rich and varied; it predates the rise of the Inca Empire. Many prehistoric canals are still in use; the norms and beliefs surrounding them have been evolving for centuries. Water, land, myth, religion, and the history and destiny of the peoples of the Sierra have been inextricably bound (Sherbondy, 1985). Layered upon this Andean tradition are the legal and social institutions governing water brought by the Spanish conquerors—e.g., rotation, the concept of water as property of the state, and the water judge. Finally, with the interjection of the state into irrigation development during this century, a third set of norms and institutions entered to govern participation. These are based on the policies and economic goals of both Peruvian government and international donors.

The present national strategy for small-scale irrigation development assumes local participation in construction activities. It also assumes that upon project completion, irrigators will organize to assume responsibility for system operation, maintenance, and repair. Given the scarce resources available for irrigation development and management in this debt-burdened nation and many widely scattered small sites in the Sierra, these assumptions regarding local participation are necessary if works are to be constructed and systems managed. Yet despite a vigorous tradition of participation in irrigation in the Andes, labor contributions to agency-assisted efforts are often smaller than anticipated and local irrigation organizations frequently fail to function.

This chapter briefly reviews the nature of participation in Andean irrigation projects, outlines factors which may adversely affect local participation, and offers recommendations for strengthening local irrigation organizations.

The Nature of Participation

Phases of Project Development and Irrigation Activities

Participation may occur in distinct phases of irrigation development: initiation, planning, construction, operation and maintenance, and rehabilitation. Within these phases, participation is focused on water, structures, and organization (Uphoff, 1986a). Management activities in which participation may take place include water acquisition, allocation, distribution, and drainage; structure design, construction, and O&M; and, at the organizational level, decision-making, resource mobilization and management, communication, and conflict resolution. In small, community-managed Andean systems, farmers typically engage in all these activities and in all phases of project development. In agency-managed systems, local participation is usually restricted to the last three phases and to activities related to system construction and O&M, water distribution, and conflict resolution.

Control and contributions. It is useful to distinguish between two forms of participation: control and contributions. In general, agencies assume that participation will consist largely of a labor contribution first during construction and later for annual cleaning and periodic maintenance and repair, but a major determinant of this contribution in the Andes is the configuration of opportunities for participation in all phases of project development and the degree to which those who are asked to contribute their labor and resources to a system also participate in decision-making and exercise control.

The labor contribution. Labor mobilization for public works in the Andes is rooted in several distinct traditions: reciprocal obligations among households, communal self-help, hacienda obligations, and *corvée* labor (obligatory, but unremunerated service building or repairing roads, canals, bridges, etc., for the state). These traditions are governed by two opposed norms: (a) reciprocity--a symmetrical norm--involving the interdependence characteristic of many interactions in relatively undifferentiated communities; and (b) the asymmetrical norm of subservience and obligation to state or landlord, deriving from the Incaic *mit'a* and from the colonial institutions of *encomienda*. Following independence, these traditions were preserved as *la República* labor.

The *mit'a* was essentially an obligation to provide goods or services to the state imposed by the Incas on subject populations. It often took the form of *corvée* labor on roads, canals, and other infrastructure. *Encomienda* was the entitlement to the use of labor of and extraction of tribute from a specific indigenous population granted to an individual colonist in return for his nominal obligation to Christianize the population. *Repartimiento* consisted of an obligation of indigenous settlements to provide a certain percentage of its population to serve limited terms as workers in agricultural, mining, or other enterprises.

A major use of the institution of *la Repbúlica* labor in Peru was the Road Conscription Law of 1921 which required able-bodied men between the ages of 21 and 50 to spend twelve days a year constructing automobile roads. In return, municipalities were expected to provide food and drink where customary and to pay a minimum wage. The unpopular law contributed to the overthrow of the Leguia government (Dobyns & Doughty, 1976).

Labor obligations to landlords usually included canal maintenance. While resident hacienda workers may have derived some benefit from canal maintenance, the primary beneficiary of this obligation was the landlord.

In the Peruvian Sierra, local and provincial elites have in the past appealed to the norms of reciprocal obligations and community self-help in order to mobilize labor for projects with private or at best very narrow ends (e.g. provision of electric power to the town center, improvement of roads serving large estates). The distinction between reciprocal and asymmetrical labor contributions becomes muddled when the ideology of reciprocal labor exchange is invoked to ensure local participation in public works projects designed for the benefit of a few. A common consequence is increased skepticism on the part of local farmers asked to contribute their labor to public works projects.

Integrating contributions and control. Until the second half of this century, virtually all irrigation systems in Peru were initiated, planned, and constructed either by *hacienda* managers or by a collectivity of water users. Those who worked the land—whether as *resident hacienda workers*, sharecroppers, or independent smallholders--contributed labor and other resources to system construction, maintenance, and rehabilitation. In the communal systems, contributions and decision-making were integrated. Rights to irrigation waters implied physical and spiritual custodianship of the system (Mitchell, 1976; Isbell, 1978). In Cuzco, participation in O&M affirms one's ancestral claims both to water rights and to community membership (Sherbondy, 1985). In contrast, control over *hacienda* systems was usually delegated by the owner to an overseer while contributions were made by the resident labor force.

Labor irrigation organization. When the Peruvian state entered into irrigation development in the Sierra, it assumed that local participation in system construction and in O&M would continue, but through new organizational forms. In some systems this meant the replacement of water judges—men of prestige and authority elected to office by the water users; in others, the separation of water management and other administrative functions heretofore combined in the *varayoc* or in the District Council. (The *varoyoc* was the council representing the indigenous population and was responsible for management of community lands. District Councils represented the *vecinos* or townsmen, not defined as *Indian* by the census.) Peruvian water law now requires that participation in agency-developed systems be channeled through local irrigation committees, *Comites de Regantes*. These organizations are intended to (1) facilitate agency-farmer interactions, (2) facilitate and organize local contributions, and (3) assume responsibility for system O&M once a project is complete.

In some areas, the *Comite* is the lowest tier in a series of water user organizations, which includes the *Comisión de Regantes* at the system level, and a *Junta de Usuarios* representing different types of water users either at a river basin or provincial level. The *Comités* are authorized to levy fees to cover the cost of materials and administration, but because water rights in the Sierra were generally allocated to users prior to system improvements, the *Comités* often find it difficult to justify to water users the imposition of charges.

Factors Affecting Participation

A number of economic, social, and political factors may stimulate or discourage participation in meetings and in construction and maintenance activities. These include the role of irrigated agriculture in the household economy, the timing of agricultural activities, and the distribution of land and wealth in the command area, demographic phenomena and local organizational capacity, agency behavior— rights in infrastructure, involvement of farmers in phases of project development, empowerment of local organizations, relationship of contributions to control.

The Role of Irrigation in the Household Economy

Subsistence vs. market production. Whether irrigation is a stimulus to market or subsistence production depends on the household's ability to meet its subsistence needs and on such factors as prices, transport, risk, etc. In a few regions with good access to national and international markets, Sierra farmers may produce irrigated crops for export and national markets.

Productive activities in time and space. In the harsh Andean environment, rural households engage in a complex of agricultural and non-agricultural activities dispersed in space and carefully integrated in time. Spatial dispersal means that household members-whether herding at high altitudes, holding urban jobs, or migrating seasonally to other ecological zones—are often working at some remove from the irrigation systems for which they are responsible. Elasticity in the timing of these activities is limited, and, because irrigation permits flexibility, tasks on lower irrigated lands will tend to be performed as time permits, while dryland activities will be more carefully scheduled due to limitations of rain and frost (Golte, 1980).

The need for households to disperse their productive activities in time and space limits the possibilities for participation in irrigation management and reduces the labor force available for system construction or O&M. Collins (1986) notes that as Puno area farmers have become more dependent upon lowland coffee production, their ability and willingness to participate in irrigation related rituals and canal cleaning has declined. In one Plan MERIS subproject near Cajamarca, O&M and construction activities have been restricted to weekends, when part-time cultivators have time for agricultural activities. While irrigation can potentially augment small farmers' food supply and permit cash crop production, this group must carefully weigh the opportunity costs of labor contributions and the risks of irrigated agriculture against potential benefits in a context of limited access to markets and credit.

Land tenure and land-labor relationships. Ability to participate and forms of participation are often determined by access to land. Large landholders usually have disproportionately greater access to cash in Peru. In the Central Sierra, they pay for food and entertainment for work parties rather than participate directly (Mallon,

1983; Walter, 1983). In Cajamarca, they send day laborers or share croppers to perform labor obligations. The latter are often smallholders within the same irrigation system (owning between 0.01 and about 3.5 hectares) who are too poor to translate their own labor obligation into a cash payment and who have little if any time left from their other productive activities to spend on irrigation. This group derives over half of its income from wage labor both inside and outside of agriculture and less than 25 percent from agriculture in Cajamarca and as little as 14 percent in the Southern highlands (Caballero, 1980; Collins, 1986; Figueroa, 1984; Lynch, 1988).

Poor farmers may work on a project, not in anticipation of future benefits, but because it provides *immediate* income in the form of food assistance. This has had adverse impacts on project quality and completion time. For example, absenteeism was a major problem for a small improvement project in La Rinconada, a small Puno hamlet; the labor force consisted largely of adolescents and women with young children and contained few project beneficiaries. Workers came mainly from neighboring communities and received payment partly in the form of food assistance and the remainder as minimum wage paid by local irrigators. (Puno data was gathered in the course of a preliminary investigation for the WMS-II Small-Scale Irrigation Systems Special Study conducted by Cornell University.) The workers stood to gain nothing from the project and found the wage unattractive; local residents, on the other hand, transformed their labor obligation into a cash and food payment rather than invest their own time in system improvement. In this case, labor was performed by a group which viewed the project as a source of immediate income rather than as a collective investment.

The Social Environment

Demographic impacts. These demographic factors may have an adverse impact on the sustainability of locally managed systems: increased population pressure on irrigated lands, outmigration of active males, and increased heterogeneity. In the Collini small-scale system in Puno and the Choloque Canal system in Cajamarca, increasing pressure on land created stresses on preexisting and functioning local systems that were relieved through the introduction of a water schedule by the local *Dirección de Aguas* officer (*sectorista*). Heavy outmigration of active males from sierra communities reduces the population available for canal cleaning and maintenance has placed additional organizational and physical burdens on resident women (Deere, 1977; Deere, 1978; Weil, 1983).

Finally, concentration of population in newly irrigated zones may introduce an element of cultural heterogeneity which reduces local organizational capacity and undermines traditional institutions. *Legitimacy, authority, and leadership.* As the nature of Sierra integration into the national economy and society changes, so do relationships between local irrigation institutions and their constituents. The legitimacy, and hence authority of water judges and the *varayoc* over irrigation matters has been challenged by the creation of *Comités de Regantes.* In turn, the viability of the *Comités* has been reduced when irrigation agencies seek to coopt or circumvent them. These phenomena have a debilitating effect on local participation and organizational capacity.

Conversely, agency involvement in local irrigation allows for the development of a new cadre of leaders, who are able to give voice to local demands within the context of the political system. These leaders often play a necessary brokerage role between agency and community. But, they are effective only to the extent that they enjoy legitimacy in the eyes of water users (Lynch, 1991).

History of Water Management and Agency Behavior

Rights in infrastructure. As agencies become involved in irrigation regulation and development in the highlands, the structure or property rights in water and irrigation infrastructure has changed. As noted above, rights in land and water were traditionally inseparable and attached to the peoples whose ancestors bore a mythic relationship to the landscape in which they lived. Participation in canal cleaning was a privilege insofar as it was an assertion of an ancestral right (Sherbondy, 1985). After independence, privately built systems were considered the private property of the landowner(s). According to present Peruvian water law, both water and infrastructure in agency-built or rehabilitated systems belong to the state. Because irrigators are no longer owners of the infrastructure in the latter systems, they are less inclined to assume responsibility for maintenance and repairs.

Relationship of Contributions to Control

The local level. The principle underlying labor mobilization for communal self-help projects in the Andes is egalitarian in theory: beneficiary households are required to furnish an equitable labor contribution—usually in proportion to number of irrigable plots, rather than hectares. This places a heavier burden on farmers with small holdings. But even where contributions are directly proportional to water allocated, the burdens of labor contributions tend to fall more heavily upon poor families.

Control is also theoretically shared equally among all irrigators; however, in highly stratified communities, large landholders tend to hold power positions or to ignore regulations of the *Comités* that run counter to their interests. Where women do a disproportionate share of the irrigating, they still tend to be underrepresented on *Comités*. Finally, where Plan MERIS made a concerted effort to include resident farmers rather than absentee landlords on the *Comité de Regantes*, these *Comités* found themselves unable to exercise control over water management (Nunberg, 1983). In general, the more cohesive and less stratified the community, the more likely it is that contributions and control will be distributed evenly.

The agency and control. Some externally funded projects are responses to local requests, but many are initiated in the absence of local initiative, and with little regard to the perceived needs of the potential beneficiaries. The Asillo project in Puno, completed in 1959, was undertaken with no local input. As a result, it was underutilized and resented by presumed beneficiaries who objected to the imposition of fees and to the design of the system (Castillo, Castillo, & Revilla, 1969). The Plan MERIS subprojects in general have suffered from low levels of farmer involvement in planning and design (Nunberg, 1983). In San Marcos, where farmers played a role in project initiation, their labor contributions and interest in water management were sustained at far higher levels than where irrigation development had no local backing.

Conclusions and Recommendations

What then are the implications of the Peruvian case for small-scale irrigation development? Irrigation planners must recognize that even where community self-help institutions are strong, they cannot be taken for granted. The quality of the labor contribution will depend upon the perceived needs of potential beneficiaries, the degree to which workers expect their contribution will affect their ultimate benefits from the project, the burden posed by a contribution on individual families, and the history of public works projects in the region.

Irrigation agencies can increase the likelihood that potential irrigators will view their contributions as worthwhile investments. A first step is to ascertain local support for the irrigation project by (1) responding to local requests for assistance, and (2) contacting a broad spectrum of residents within the community to identify groups promoting irrigation development.

Second, a baseline study should examine local productive strategies and land tenure relationships in order to gauge the extent and quality of participation. Attention to external demands on irrigators' time in agriculture and other productive activities will enable an agency to schedule construction for periods when labor is relatively abundant. This implies that materials and equipment will arrive on site in a timely manner, which has been the exception rather than the rule in Peru.

Finally, the agency must take steps to assure participants that their labor contributions will yield a share in control over the system by involving them in the earliest stages of project planning and by promoting the development or recognition of water user associations accountable to the community as well as to the agency. Participation in O&M is likely to be greater where irrigators not only control decision making, but are owners of the systems for which they are made responsible.

References

Abeyratne, S., Brewer, J., Ganewatte, P., & Uphoff, N. (1984). *Improving irrigation management through farmer organization: Responses to a program in Sri Lanka.* Paper prepared for SSRC South Asia Committee Conference on Community Responses to Irrigation, Bangalore, India.

Adams, A. (1981). The Senegal River Valley. In J. Heyer, P. Roberts, & G. Williams (Eds.), *Rural development in tropical Africa.* New York: St. Martin's.

Adams, N. L. (1983). *Synthesis of lessons learned for rapid appraisal of irrigation strategies* (Water Management Synthesis Project Report No. 22). Logan, UT: Utah State University.

Adams, W. M. (1984). Irrigation as hazard: Farmers' responses to the introduction of irrigation in Sokoto, Nigeria. In W. M. Adams & A. T. Grove (Eds.), *Irrigation in tropical Africa.* Cambridge: University of Cambridge, African Studies Centre.

Alwis, J. (1983). *System H of the Mahaweli development project, Sri Lanka: 1982 diagnostic analysis* (Water Management Synthesis Project Report No. 16). Ft. Collins, CO: Colorado State University.

Anderson, M., & Gehrke, C. (1987, Fall). The degradation factor. *Wilderness,* pp. 38-40.

Andrae, G., & Beckman, B. (1985). *The wheat trap, bread and underdevelopment in Nigeria.* London: Zed Books.

Anna University, Madras. (1982). *Proceedings of the International Workshop on Modernization of Tank Irrigation System - Problems and Issues.* Madras, India: Anna University.

Agrarian Research Training Institute. (1985). *The institutional mechanism for irrigation water management in the Gal Oya Left Bank and the periphery: A case study of the role of the wattei vidhane.* Paper prepared as a sub-study under the Water Management Project, Colombo, Sri Lanka.

Axelrod, R. (1984). *The evolution of cooperation.* New York: Basic Books.

Baden, J. (1977). A primer for the management of common pool resources. In G. Hardin & J. Baden (Eds.), *Managing the Commons* (pp. 137-146). San Francisco: W. H. Freeman & Co.

Bagadion, B. U. (1985). Farmers' involvement in irrigation development in the Philippines. *In Participatory experiences in irrigation water management* (pp. 61-76). Rome: Food and Agriculture Organization of the United Nations.

———. (1988). *Farmer participation in irrigation management in the Philippines.* West Hartford, CT: Kumarian Press.

———., & Korten, F. K. (1985). Developing irrigator's organizations: A learning process approach to participatory irrigation programs. In M. Cernea (Ed.), *Putting people first: Sociological variables in rural development* (pp. 52-90). World Bank, New York: Oxford University Press.

Banfield, E. C. (1958). *The moral basis of a backward society.* New York: Free Press.

Barclay, H. B. (1986, August). *Segmental acephalous network systems: Alternatives to centralized bureaucracy.* Paper presented at the World Congress of Sociology, New Delhi, India.

Barker, R., Coward, E. W., Levine, G., & Small, L. E. (1984). *Irrigation development in Asia: Past trends and future directions* (Cornell University Studies in Irrigation No. 1). Ithaca, NY: Cornell University.

Barkley, J. R. (1981). *Northern Colorado Water Conservancy District.* Loveland, CO: Colorado Big Thompson Project.

Barnett, T. (1977). *The Gezira scheme: An illusion of development.* London: Frank Cass.

Baum, G., & Migot-Adholla, S. (1982). South Kano irrigation scheme: Socio-economic adjustments in rural development planning. *Quarterly Journal of International Agriculture, 21*(1), 37-51.

Bautista, H. B. (1987). *Experiences with organizing irrigators associations: A case study from the Magat River irrigation project in the Philippines.* Kandy, Sri Lanka: International Irrigation Management Institute.

Begum, S. (1985). *Minor tank water management in the dry zone.* Paper prepared for the Agrarian Research and Training Institute, Colombo, Sri Lanka.

Bhandarkar, V., & Freeman, D. M. (1988). A tank system in Madhya Pradesh, India. Vol. 3 in: *Linking main and farm irrigation systems in order to control water* (Water Management Synthesis Project Report 69). Colorado State University, Fort Collins, CO.

Bingen, R. J. (1985). *Food production and rural development in the Sahel:Lessons from Mali's operation Riz-Segou.* Boulder, CO: Westview Press.

Blackie, M. J. (Ed.). (1984). *African regional symposium on small-holder irrigation.*Harare, Zimbabwe: University of Zimbabwe.

Blair, D. H., & Pollack, R. A. (1983). Rational collective choice. *Scientific American, 249*(2), 88-95.

Bloch, P. (1986). Senegal: Senegal River Basin. In P. Bloch (Ed.), *Land tenure issues in river basin development in sub-Saharan Africa* (pp. 29-41). Madison, WI: University of Wisconsin-Madison, Land Tenure Center.

Blomquist, W., & Ostrom, E. (1985). Institutional capacity and the resolution of a commons dilemma. *Policy Studies Review, 5*(2), 383-393.

Blumberg, P. (1969). *The sociology of participation.* New York: Schoken Books.

Borlaug, N. E. (1987). Making institutions work--a scientist's viewpoint. In W. R. Jordan (Ed.), *Water and water policy in world food supplies* (pp. 387-396). College Station, TX: Texas A&M University Press.

Boslough, J. (1981, June). Rationing a river. *Science 81,* pp. 26-38.

Bottrall, A. F. (1978). The management and operation of irrigation schemes in less developed countries. *Water Supply and Management, 2.*

_____ . (1981a). Improving canal management: The role of evaluation and action research. *Water Supply and Management, 5*(1).

_____ . (1981b). *Comparative study of the management and organization of irrigation projects* (World Bank Staff Working Paper #458). Washington, DC: World Bank.

_____ . (1985). *Managing large irrigation schemes: A problem of political economy* (Occasional Paper #7). London: Agricultural Administration Unit, Overseas Development Institute.

Bray, F. (1986). *The rice economies: Technology and development in Asian societies.* New York: Basil Blackwell.

Bromley, D. W. (1982). *Improving irrigated agriculture: Institutional reform and the small farmer* (World Bank Staff Working Paper No. 531). Washington, DC: The World Bank.

_____ . (1987). Irrigation institutions: The myth of management. In W. R. Jordan (Ed.), *Water and water policy in world food supplies* (pp. 173-176). College Station, TX: Texas A&M University Press.

Bryant, C., & White, L. (1984). *Managing rural development with small farmer participation.* West Hartford, CT: Kumarian Press.

Buchanan, J. M., & Tullock, G. (1962). *The calculus of consent: Logical foundations of constitutional democracy.* Ann Arbor, MI: University of Michigan Press.

Bullock, K., & Baden, J. (1977). Communes and the logic of the commons. In G. Hardin & J. Baden (Eds.), *Managing the Commons* (pp. 182-199). San Francisco: W. H. Freeman & Co.

Byrnes, K. (1985). *The potential role of farmer organizations in increasing the productivity and income earning capability of small-farmer agricultural systems in the developing countries.* Washington, DC: Academy for Educational Development.

_____ . (1986). *Using farmer organizations to support communication for technology transfer in agriculture.* Washington, DC: Academy for Educational Development.

Caballero, J. M. (1980). *Agricultura, reforma agraria, y pobreza compesina.* Lima: Instituto de Estudios Peruanos.

Campbell, D. (1986, November). Central Arizona Project: The faucet opens. *Arizona Highways,* pp. 6-11.

Carruthers, I. (1983). *Aid for the development of irrigation.* Paris: Organization for Economic Cooperation and Development.

Castillo, H., Castillo, T., & Revilla, A. (1969). *Accopata: The reluctant recipient of technological change* (Socioeconomic Development of Andean Communities, Report No. 2). Cornell-Peru Project, Department of Anthropology, Cornell University.

Cernea, M. (1983). *A social methodology for community participation in local investments: The expense of Mexico's PIDER program* (World Bank Staff Working Paper #598). Washington, DC: World Bank.

_____ . (Ed.). (1985). *Putting people first.* New York: Oxford University Press.

Chakravarty, T. K., & Das, P. K. (1982). Community minor irrigation scheme in West Bengal: A case study. *Journal of Rural Development,* 1(3),445-457.

Chambers, R. (1980a). Basic concepts in the organization of irrigation. In E. W. Coward, Jr. (Ed.), *Irrigation and agricultural development in Asia* (pp. 28-50). Ithaca, NY: Cornell University Press.

_____ . (1987). Food and water as if poor people mattered: A professional revolution. In W. R. Jordan (Ed.), *Water and water policy in world food supplies* (pp. 15-21). College Station, TX: Texas A&M University Press.

_____ ., & Moris, J. (1973). *Mwea: An irrigated rice settlement in Kenya.* Munich: IFO-Institut, Weltforum Verlag.

Champion, D. (1975). *The sociology of organizations.* New York: McGraw Hill.

Coleman, J. S. (1990). *Foundations of social theory.* Cambridge: Harvard University Press.

Collins, J. L. (1986). *Labor dynamics, producer decisions, and cycles of environmental decline.* Paper revised for Lands at Risk in the Third World: Local Level Perspectives.

Coser, L. (1956). *The functions of social conflict.* New York: The Free Press.

Coward, E. W., Jr. (1980). Irrigation development: Institutional and organizational issues. In E. W. Coward, Jr. (Ed.), *Irrigation and Agricultural Development in Asia* (pp. 15-27). Ithaca, NY: Cornell University Press.

_____ ., Jr. (1986a). State and locality in Asian irrigation development: The property factor. In K. E. Nobe & R. K. Sampath (Eds.), *Irrigation management in developing countries: Current issues and approaches* (pp. 491-508). Boulder, CO: Westview Press.

_____ ., Jr. (1986b). Direct or indirect alternatives for irrigation investment and the creation of property. In K. W. Easter (Ed.), *Irrigation investment, technology, and management strategies for development* (pp. 225-244). Boulder, CO: Westview Press.

_____ ., Jr. (1987). Action experiments in irrigation development: Identifying policy and program implications. In W. R. Jordan (Ed.), *Water and water policy in world food supplies* (pp. 361-366). College Station, TX: Texas A&M University Press.

Craven, K., & Tuluy, A. H. (1981). Rice policy in: Senegal. In S. Pearson, J. Stryker, & C. Humphreys (Eds.), *Rice in west Africa: Policy and economics* (pp. 229-262). Stanford, CA: Stanford University Press.

Datye, K. R., & Patil, R. K. (1987). *Farmer managed irrigation systems.* Bombay, India: Centre for Applied Systems Analysis in Development.

de los Reyes, R. P. (1980). *Managing communal gravity systems.* Quezon City: Institute of Philippine Culture.

de Silva, N. G. R. (1981). Farmer participation in water management: The Minipe project in Sri Lanka. *Rural Development Participation Review,* 3(1),16-19.

De Wilde, J., McLoughlin, P., Gunard, A., Scudder, T., & Maubouché, R. (1967). *Experiences with agricultural development in tropical Africa* (Vol. 2). Baltimore, MD: Johns Hopkins Press.

Deere, C. D. (1977). Changing social relations of production and Peruvian peasant women's work. *Latin American Perspectives*, 4, 48-69.
_____. (1978). *The development of capitalism in agriculture and the division of labor by sex: A study of the northern Peruvian Sierra.* Unpublished doctoral dissertation, University of California, Berkeley.

Deshpande, V. D. (1986). Pani Panchayat: A movement to secure irrigation for poor farmers. *Symposium on Community Management in Irrigation Systems* (pp. 1-26). Pune, India: National Institute of Bank Management.

Dey, J. (1984). *Women in rice-farming systems: Focus: sub-Saharan Africa.* (Women in Agriculture, Vol. 2). Rome: Food and Agriculture Organization of the United Nations.

Dobyns, H. E., & Doughty, P. L. (1976). *Peru: A cultural history.* New York: Oxford University Press.

Doorenbos, J., & Kassam, A. H. (1979). *Yield response to water.* Rome: Food and Agricultural Organization of the United Nations.

Drucker, P. (1974). *Management.* London: Heineman.

Dunbar, R. (1983). *Forging new rights in western waters.* Lincoln: University of Nebraska Press.

Easter, K. W., & Welsch, D. E. (1986). Priorities for irrigation planning and investment, and implementing irrigation projects: Operational and institutional problems. In K. W. Easter (Ed.), *Irrigation management, technology, and management strategies for development* (pp. 14-56). Boulder, CO: Westview.

El-Ashry, M., & Gibbons, D. (1988). The West in profile. In M. El-Ashry & D. Gibbons (Eds.), *Water and arid lands of the western United States* (pp. 1-19). Cambridge: Cambridge University Press.

Engineering News. (1985, January). Big irrigation projects fail at feeding Africa. *Engineering News*, pp. 10.

Esman, M. J., & Uphoff, N. (1984). *Local organizations: Intermediaries in rural development.* Ithaca, NY: Cornell University Press.

Faki, H. (1982). Disparities in the management of resources between farm and national levels in irrigation projects, example of the Sudan Gezira scheme. *Agricultural Administration*, 9(1), 47-59.

Farmer, B. H. (1957). *Pioneer peasant colonization.* Oxford, England: Oxford University Press.

Figueroa, A. (1984). *Capitalist development and the peasant economy in Peru.* Cambridge, England: Cambridge University Press.

Folk-Williams, J., Fry, S., & Hilgendorf, L. (1985). *Water in the West: Western water flows to the cities.* (Vol. 3). Santa Fe, NM.: Western Network.

Food and Agriculture Organization of the United Nations. (1985). Participatory experiences in irrigation water management. *Proceedings of the Expert Consultation on Irrigation Water Management.* Rome: Food and Agriculture Organization of the United Nations.
_____. (1986a). *African agriculture: The next 25 years* (Main Report). Rome: Food and Agriculture Organization of the United Nations.

242

_____ . (1986b). *African agriculture: The next 25 Years* (Annex IV, Irrigation and Water Control). Rome: Food and Agriculture Organization of the United Nations.

_____ . (1986c). *Irrigation in Africa south of the Sahara* (Food and Agriculture Organization of the United Nations Investment Centre Technical Paper, No. 5). Rome: Food and Agriculture Organization of the United Nations.

Foster, G. M. (1973). *Traditional societies and technological change*. New York: Harper & Row.

Foster-Carter, A. (1985). *The sociology of development*. Ormskirk, England: Causeway Press, Ltd.

Francis, P. (1984). For the use and common benefit of all Nigerians: Consequences of the 1978 land nationalization. *Africa*, 54(3), 5-28.

Frank, A. G. (1969). *Capitalism and underdevelopment in Latin America: Historical studies of Chile and Brazil*. New York: Monthly Review Press.

Frederick, K., & Hanson, J. (1982). *Water for Western agriculture*. Washington, D.C.: Resources for the Future.

Freeman, D. M. (1986, August). *A comparative sociology of irrigation systems: Concepts and variables*. Paper presented at the Rural Sociological Society Meetings, Salt Lake City, UT.

_____ . (1988). *Designing local organizations for reconciling water supply and demand. Linking main and farm irrigation systems in order to control water* (Water Management Synthesis Project Report 69, Vol. 1 of 5). Fort Collins, CO: Colorado State University.

_____ . (1989). *Local organizations for social development: Concepts and cases of irrigation development*. Boulder, CO: Westview Press.

_____ . (1990). Designing local irrigation organizations for linking water demand with supply. In R. K. Sampath & R.W. Young (Eds.), *Social, economic, and institutional issues in third world irrigation management* (pp. 111-140). Boulder, CO: Westview.

_____ ., & Lowdermilk, M. L. (1981). Sociological analysis of irrigation water management: A perspective and approach to assist decisionmaking. In C. S. Russell & N. K. Nicholson (Eds.), *Public choice and rural development* (pp. 153-173). Washington, DC: Resources for the Future.

_____ ., & Lowdermilk, M. L. (1985). Middle-level organizational linkages in irrigation projects. In M. Cernea (Ed.), *Putting people first-sociological variables in rural development* (pp. 91-118). New York: Oxford University Press.

Fresson, S., Amselle, J., Bagayoko, D., Benhamou, J., Leuilier, J., & Ruf, T. (1985). *Evaluation de l'office du Niger (Mali)*. Paris: Ministere des Relations Exterieures Cooperation et Developement.

Frohock, F. M. (1987). *Rational association*. Syracuse, NY: Syracuse University Press.

Frolick, N. & Oppenheimer, J. (1970). I get by with a little help from my friends. *World Politics*, 23:104-120.

Gaitskell, A. (1959). *Gezira: A story of development in the Sudan*. London: Faber & Faber.

Ganewatte, P. (1985). *Farmer participation in planning, construction and management of irrigation systems*. Paper prepared for the International Irrigation Management Institute Course on Planning and Management of Irrigation Systems, Digana, Sri Lanka.

Gichuki, F. (1985, October). *Small-scale irrigation in Kenya*. Water Management Synthesis II Seminar, Utah State University, Logan, UT.

Gitonga, S. (1985). *Bura irrigation and settlement project, Kenya*. Paper for the Informal Meeting on Irrigation in Anglophone Africa, Food and Agriculture Organization of the United Nations, Rome.

Goetze, D. (1986). Identifying appropriate institutions for efficient use of common pools. *Natural Resources Journal, 27*(1), 187-200.

Goldsmith, E., & Hildyard, N. (1984). *The social and environmental effects of large dams*. San Francisco: Sierra Club.

_____., & Hildyard, N. (1986, May/June). Dams: They are destroying habitat and wildlife along many of the world's rivers. *Defenders*, pp. 19-33.

Golte, J. (1980). *La racionalidad de la organizacion Andina*. Lima: Instituto de Estudios Peruanos.

Gottlieb, R. (1988). *A life of its own: The politics and power of water*. New York: Harcourt Brace Jovanovich.

Government of Pakistan. (1984). *Water users associations in Pakistan*. Islamabad: Ministry of Food, Agriculture and Cooperatives.

Gross, E., & Etzioni, A. (1985). *Organizations in society*. Englewood Cliffs, NJ: Prentice Hall.

Grusky, O., & Miller, G. A. (1981). *The sociology of organizations*. New York: The Free Press.

Gunasekera, W. (1982). The role of traditional water management in modern paddy cultivation in Sri Lanka. *Marga Quarterly, 6*(3), 69-123.

Haas, E. J., & Drabek, T. E. (1973). *Complex organizations: A sociological perspective*. New York: The MacMillan Co.

Hagen, E. (1968). *The economics of development*. Homewood, IL: Richard D. Irwin, Inc.

Haider, M. I., Sheng, T. S., & Tinsely, R. L. (1986). *Pre-rehabilitation diagnostic study of Sehra irrigation system, Sind, Pakistan*. Fort Collins, CO: Colorado State University, Water Management Synthesis Project.

Hall, R. (1991). *Organizations: Structures, processes and outcomes*. Englewood Cliffs, NJ: Prentice Hall.

Hameed, N. D. A. (1977). *Rice revolution in Sri Lanka*. Geneva: United Nations Research Institute for Social Development.

Harmon, M. M., & Mayer, R. T. (1986). *Organization theory for public administration*. Boston, MA: Little, Brown & Co.

Harrison, L. E. (1985). *Underdevelopment is a state of mind: The Latin American case*. Lanham, MD: University Press of America.

244

Hart, H. (1978). Anarchy, paternalism, or collective responsibility under the canals. *Economic and Political Weekly, 13*, 51-52.

Hawley, A. H. (1973). Human ecology. In M. Micklin (Ed.), *Population, environment and social organization: Current issues in human ecology* (pp. 27-43). Hinsdale, IL: Dryden Press.

Herring, R. J. (1985). *Food policy and welfare in Sri Lanka: Before and after the liberalization regime.* Paper presented at the Conference on the Political Economy of Food, Utah State University, Logan, UT.

Heuvelmans, M. (1974). *The river killers.* Harrisburg, PA: Stackpole Books.

Hirschman, A. (1963). *Journey towards progress.* New York: Twentieth Century Fund.

Hogg, R. (1983). Irrigation agriculture and pastoral development: A lesson from Kenya. *Development and Change, 14*, 577-599.

Hunt, R., & Hunt, E. (1976). Canal irrigation and local social organization. *Current Anthropology, 17*(3).

Illo, J. F., & Chiong-Javier, M. E. (1983). *Organizing farmers for irrigation management: The Buhi-Lalo experience.* Naga City, Philippines: Research and Service Center.

India Famine Commission. 1898.

_____ . 1901.

Isbell, B. J. (1978). *To defend ourselves: Ecology and ritual in an Andean village* (Latin American Monographs No. 47). Institute of Latin American Studies, University of Texas at Austin.

Jagawat, H. (1986). *Lift irrigation cooperative: An experience in Panchamahal.* Paper presented at Symposium on Community Management in Irrigation Systems, National Institute of Bank Management, Pune, India.

Jain L. C., Krishnamurthy, B. V., & Tripathi, P. M. (1985). *Grass without roots: Rural development under government auspices.* New Delhi: Sage Publications.

James, D. (1986, April). *The Plan MERIS project in northern Peru.* Water Management Synthesis II Seminar, Utah State University, Logan, UT.

Jayawardene, J. (1984). *Water management in System H of the accelerated Mahaweli development program, Sri Lanka.* Paper presented at the Systems Management Seminar, Water Management Synthesis II Project, Fort Collins, CO.

Jayawardene, J. (1990). *Farmer and officer: Training strategies in system "H".* Accelerated Mahaweli Development Programme, Colombo, Sri Lanka.

Jesser, C. J. (1975). *Social theory revisited.* Hindsale, IL: The Dryden Press.

Kanwar, J. S. (1988). *Water management: The key to developing agriculture.* New Delhi: Indian National Science Academy.

Kathpalia, G. N. (1981). Rotational system of canal supplies and *warabundi* in India (Appendix B). In J. Keller (Ed.) India/USAID: *Irrigation development options and investment strategies for the 1980's* (Water Management Synthesis Project, Report No. 6). Logan, UT: Utah State University.

245

Keller, J. (1987). Worldwide view of irrigation technology development. In W. R. Jordan (Ed.), *Water and water policy in world food supplies* (pp. 305-311). College Station, TX: Texas A&M Press.

Keller, J. (1988). *Irrigation scheme water management* (Water Management Synthesis Project Report No. 4). Logan, UT: Utah State University.

Keller, J., Meyer, R., Peterson, F., Weaver, T., & Wheelock, G. (1982). *Project review for Bakel small irrigated perimeters* (Water Management Synthesis Project Report No. 9). Logan, UT: Utah State University.

Kelly, W. W. (1982a). *Irrigation management in Japan: A critical review of Japanese soil science research* (East Asian Papers No. 30). Ithaca, NY: Cornell University.

Kelly, W. W. (1982b). *Water control in Tokugama, Japan: Irrigation organization in a Japanese river basin, 1600-1870* (East Asian Papers No. 31). Ithaca, NY: Cornell University.

Knott, J. H., & Miller, G. J. (1987). *Reforming bureaucracy: The politics of institutional choice.* Englewood Cliffs, NJ: Prentice-Hall.

Koita, Tocka. (1986, May). *Comparison of large and small-scale irrigated perimeters of Mauritania.* Paper presented at the Irrigation Systems Research Forum, Cornell University, Ithaca, NY.

Korten, D. C. (1980). Community organization and rural development: A learning process approach. *Public Administration Review, 40*(5), 480-511.

Korten, F. F. (1982). *Building national capacity to develop water user associations: Experience from the Philippines* (Staff Working Paper No. 528). Washington, DC: World Bank.

Leach, E. R. (1961). *Pul Eliya: A village in Ceylon.* Cambridge, England: Cambridge University Press.

Levine, G., Capener, H., & Gore, P. (1972, October). *The management of irrigation systems for the farm.* Summary report presented at ADC/RTN Irrigation Seminar, Cornell University, Ithaca, NY.

Little, C. (1987, Fall). The great American aquifer. *Wilderness,* pp. 43-47.

Longhurst, R. (Ed.). (1986). Seasonality and poverty [special issue]. *IDS Bulletin, 17*(3).

Lovato, P. (1974). *Las acequias del Norte* (Technical Report No. 1). Taos, NM: Four Corners Regional Commission.

Lowdermilk, M. (1985). *Improved irrigation management: Why involve farmers?* (Network Paper No. 11c, May). London: Overseas Development Institute.

Lowdermilk, M. K. (1986). Improved irrigation management: Why involve farmers? In K. C. Nobe & R. K. Sampath (Eds.), *Irrigation management in developing countries: Current issues and approaches* (pp. 427-456). Boulder, CO: Westview Press.

_____., Early, A. C., & Freeman, D. M. (1978). *Farm irrigation constraints and farmers' responses: Comprehensive field survey in Pakistan* (Water Management Synthesis Technical Report No. 48; A-F). Fort Collins, CO: Colorado State University.

Lusk, M. W., & Parlin, B. W. (1986, August). *Overcoming social obstacles to irrigation organization.* Paper presented at the World Congress of Sociology, New Delhi, India.

———., & Parlin, B. W. (1988). *Bureaucratic and farmer participation in irrigation development* (Water Management Synthesis Report 89). Logan, UT: Utah State University.

———., & Riley, P. J. (1986). Public choice theory and irrigation development. *The Rural Sociologist, 6*(4), 280-289.

Lynch, B. (1985). *Community participation and local organization for small-scale irrigation* (Water Management Synthesis Report No. 34). Ithaca, NY: Cornell University Press.

Lynch, B. (1988). *The bureaucratic transistion: Peruvian government intervention in Sierra small-scale irrigation.* Unpublished doctoral dissertation, Cornell University, Ithaca, NY.

Lynch, B. (in press). Women and irrigation in highland Peru. *Society and Natural Resources.*

———. (1986). *Plan piloto final report: Sociological investigations* (Water Management Synthesis II Project). Ithaca, NY: Cornell University.

Maass, A., & Anderson, R. L. (1986). *And the desert shall rejoice: Conflict, growth, and justice in arid environments.* Malabar, FL: Robert E. Krieger Publishing Co., Inc.

Mallon, F. (1983). *The defense of community in Peru's central highlands.* Princeton, NJ: Princeton University Press.

Maltby, E. (1985). *Waterlogged wealth: Why waste the world's wet places?* London: Earthscan.

Martin, E., & Yoder, R. (1983). *Water allocation and resource mobilization for irrigation: A comparison of two systems in Nepal.* Paper presented at the meeting of The Nepal Studies Association, University of Wisconsin, Madison, WI.

Mathur, J. R. (1984). *Water resources: Distribution, use and management.* New York: John Wiley & Sons.

Maurya, P. R., & Sachan, R. S. (1984). Large and small-scale irrigation systems in Nigeria: A comparative study. In M. J. Blackie (Ed.), *African regional symposium on small-holder irrigation.* Harare, Zimbabwe: University of Zimbabwe.

McConnen, R. (1986, May). *Research on irrigation in Africa: An overview.* Paper presented at the Irrigation Systems Research Forum, Cornell University, Ithaca, NY.

McGuire, M. F., & Ruttan, V. W. (1990). Lost directions: U.S. Foreign Assistance Policy since New Directions. *The Journal of Developing Areas, 24,* 127-180.

McIntire, J. (1981). *Rice policy in Mali.* In S. Pearson, J. Stryker, & C. Humphreys (Eds.), Rice in West Africa: Policy and economics (pp. 299-360). Stanford, CA: Stanford University Press.

Medagama, J. (1984). *The village irrigation rehabilitation project*. Paper presented at the Agrarian Research and Training Institute, Colombo, Sri Lanka.

Merrey, D. J. (1986a). Reorganizing irrigation: Local level management in the Punjab (Pakistan). In D. J. Merrey & J. M. Wolf (Eds.), *Irrigation management in Pakistan: Four papers* (pp. 26-43). Kandy, Sri Lanka: International Irrigation Management Institute.

_____. (1986b). The sociology of *warabandi*: A case study from Pakistan. In D. J. Merrey & J. M. Wolf (Eds.), *Irrigation management in Pakistan: Four papers* (pp. 44-66). Kandy, Sri Lanka: International Irrigation Management Institute.

Miller, R. (1985). *Peasant autonomy and irrigation: Innovation in the Senegal river basin*. Ithaca, NY: Cornell University, Center for International Studies.

Mills, C. W. (1959). *The sociological imagination*. London: The Oxford University Press.

Mitchell, W. P. (1976). Irrigation and community in the central Peruvian highlands. *American Antiquity, 78,* 25-44.

Molden, D. J. (1987). *Water control as a basis for evaluation of irrigation water delivery systems*. Unpublished doctoral dissertation, Colorado State University, Fort Collins, CO.

Montgomery, J. D. (1974). *Technology and civic life: Making and implementing development decisions*. Cambridge, MA: MIT Press.

_____. (1983). When local participation helps. *Journal of Policy Analysis and Management, 3*(1), 90-105.

Moris, J. R. (1981) *Managing induced rural development*. Bloomington, IN: Indiana University, International Development Institute.

Moris, J. (1984). Managing irrigation in isolated environments: A case study of Action Ble'-Dire, In M. J. Blackie (Ed.), *African regional symposium on small-holder irrigation*. Harare, Zimbabwe: University of Zimbabwe

_____., & Norman, R. (1984). *Prospects for small-scale irrigation development in the Sahel* (Water Management Synthesis Report 26). Logan, UT: Utah State University.

_____., & Thom, D. (1985). *African irrigation overview* (Water Management Synthesis Project Report 37, Vols. 1-3). Logan, UT: Utah State University.

_____., & Thom, D. (1990). *Irrigation development in Africa: Lessons of experience*. Boulder, CO: Westview Press.

Mueller, D. C. (1979). *Public choice*. New York: Cambridge University Press.

Myrdal, G. (1963). *Economic theory and underdeveloped regions*. London: Methuen.

NABARD. (1982). *River life irrigation scheme in Kolhapur District* (Series No. 16). Bombay, India: NABARD.

National Water Policy. (1987). Government of India, Ministry of Water Resources, New Delhi.

Nickum, J. E. (1974). *Management of the Meichuan Reservoir Irrigation District.* San Jose, CA: San Jose State University.

Nickum, J. E. (1980). Local water management in the People's Republic of China. In E. W. Coward (Ed.), *Irrigation and agricultural development in Asia.* Ithaca, NY: Cornell University Press.

Notoatmodjo, B. (1985). Water users associations in Indonesia. In *Participatory experiences in irrigation water management.* Rome: Food and Agriculture Organization of the United Nations.

Nott, G. (1985, June). *Sri Lanka Project Evaluations* (Report to United States Agency for International Development). Colombo, Sri Lanka: United States Agency for International Development.

Nunberg, B. (1983). Appendix F: Institutional issues. In J. L. Wilkinson (Ed.), *Improved land and water use in the Sierra: United States Agency for International Development project impact evaluation.* United States Agency for International Development.

Olson, M. (1965). *The logic of collective action.* New York: Schocken Books.

Ostrom, V. (1974). *The intellectual crisis in American public administration* (rev. ed.). University, AL: University of Alabama Press.

_____., & Ostrom, E. (1975). Public goods and public choices. In E. S. Savos (Ed.), *Alternatives for delivering public services* (pp. 7-49). Boulder, CO: Westview Press.

Owens, E., & Shaw, R. (1972). *Development recommended: Bridging the gap between government and people.* Oxford: Oxford University Press.

Painter, J., Baldwin, E., & Malone, S. (1982). *The on-farm water management project in Pakistan* (Project Impact Evaluation No. 35). Washington, DC: United States Agency for International Development.

Palmer-Jones, R. (1984). Mismanaging the peasants: Some origins of low productivity on irrigation schemes in N. Nigeria. In W. M. Adams & A. T. Grove (Eds.), *Irrigation in tropical Africa.* Cambridge, England: University of Cambridge, African Studies Centre.

Pant, N. (1986). Group tubewell deterioration in Uttar Pradesh. *Symposium on community management in irrigation system.* National Institute of Bank Management, Pune, India.

_____., & Verna, R. K. (1983). *Farmers' organization and irrigation management.* New Delhi: Ashish Publishing House.

Parlin, B. W., Jayawardene, J., & Amarasena, G. (1985). *Obstacles to effective water management in a rural development scheme: The case of the Mahaweli H system.* Paper presented at the Annual Meetings of the Rural Sociological Society, Blacksburg, VA.

_____., & Lusk, M. (1988, August). International irrigation development: Factors affecting project success. *Society and Natural Resources,* 1(2), 131-144.

_____ ., Lusk, M. W., & Al-Rashid, N. (1986, August). *Irrigation development and technical transfer: Factors affecting the success of foreign aid projects.* Paper presented at the Rural Sociological Society Meetings, Salt Lake City, UT.

Parnakian, K., Laitos, W. R., & Freeman, D. M. (1988). The case of Lam Chamuak, Thailand. In D. M. Freeman (Ed.), *Linking main and farm irrigation systems in order to control water* (Water Management Synthesis Project Report 69, Vol. 4). Fort Collins: CO, Colorado State University.

Pearson, M. (1980). *Settlement of Pastoral Nomads* (Development Studies Occasional Paper No. 5.). Norwich: University of East Anglia.

Ponnambalam, S. (1980). *Dependent capitalism in crisis: The Sri Lankan economy 1948-1980.* London: Zed Press.

Postel, S. (1989). *Water for agriculture: Facing the limits.* Washington, DC: Worldwatch Institute.

Posz, G., Raj, B., & Peterson, D. F. (1981). Water resource development in India (Appendix A). In J. Keller (Ed.), *India/United States Agency for International Development: Irrigation development options and investment strategies for the 1980's* (Water Management Synthesis Project Report 6). Logan, UT: Utah State University.

Powledge, F. (1982). *Water.* New York: Farrar Straus Giroux.

_____ . (1987, Fall). The Poisoned Well. *Wilderness,* pp. 40-43.

Pradhan, P. (1984). The Chhatis Mauza irrigation system. In P. Ramamurthy (Ed.), *A report of the Conference on Community Responses to Irrigation.* Bangalore, India: Indian Institute of Management.

Radosevich, G. E. (1986). Legal and institutional aspects of irrigation water management. In K. C. Nobe & R. K. Sampath (Eds.), *Developing countries: Current issues and approaches* (pp. 457-490). Boulder, CO: Westview Press.

_____ . (1987) Water policy and law: The many links in food production. In W. R. Jordan (Ed.), *Water and water policy in world food supplies* (pp. 185-191). College Station, TX: Texas A&M University Press.

Rangeley, R. (1985, December). *Irrigation in arid Africa: Contribution and issues.* World Bank Second Irrigation and Drainage Seminar, Annapolis, MD.

Rangeley, W. R. (1987). Irrigation and drainage in the world. In W. R. Jordan (Ed.), *Water and water policy in world food supplies* (pp. 29-35). College Station, TX: Texas A&M University Press.

Raynolds, L. T. (1985). *The importance of land tenure in the distribution of benefits from irrigation development projects: Findings from the Cayes Plains, Haiti,* (Water Management Synthesis II Report). Ithaca, NY: Cornell University.

Reddy, J. M. (1986). Management of gravity flow irrigation systems. In K. C. Nobe & R. K. Sampath (Eds.), *Developing countries: Current issues and approaches* (pp. 95-115). Boulder: Westview Press.

250

Reidinger, R. B. (1974). Institutional rationing of canal water in northern India: Conflict between traditional patterns and modern needs. *Economic Development and Cultural Change*, 23.

Reisner, M. (1986). *Cadillac Desert*. New York: Viking Press.

_____., & Bates, S. (1990). *Overtapped oasis: Reform or revolution for western water*. Washington, DC: Island Press.

Report of the Indian Irrigation Commission. (1903). Calcutta: Government of India.

_____. (1972). New Delhi: Government of India.

Report of the working group on minor irrigation for formulation of the seventh plan proposals for the years 1985-90. (1984). New Delhi: Government of India.

Richards, P. (1986). *Indigenous agricultural revolution*. London: Hutchinson.

Roberts, M. (1980). Traditional customs in irrigation development in Sri Lanka. In E. W. Coward, Jr. (Ed.), *Irrigation and agricultural development in Asia: Perspectives from the social sciences* (pp. 186-202). Ithaca, NY: Cornell University Press.

Roethligberger, F. J., & Dickson, W. (1950). *Management and the worker*. Cambridge, MA: Harvard University Press.

Rostow, W. W. (1971). *The stages of economic growth: A non-communist manifesto* (2nd ed.). Cambridge, MA: Cambridge University Press.

Ruigu, G. M., Alila, P., & Chitere, P. (1984). *A social and economic survey: Bura irrigation settlement project*. Nairobi: Institute for Development Studies.

Ryan, W. (1976). *Blaming the victim* (rev. ed.). New York: Random House.

Schell, O. (1984). *To get rich is glorious: China in the 80's*. New York: Mentor Books.

Scott, R. W. (1981). *Organizations: Rational, natural and open systems*. Englewood Cliffs, NJ: Prentice Hall.

Scudder, T. (1985). *The Accelerated Mahaweli Program* (AMP) and dry zone resettlement (Report #5 to USAID). Sri Lanka: United States Agency for International Development.

Selowsky, M. (1979). *Balancing trickle down and basic needs strategies* (World Bank Staff Working Paper No. 335). Washington, DC: The World Bank.

Sengupta, N. (1984). Ahar-pyne irrigation. In P. Ramamurthy (Ed.), *A report of the Conference on Community Response to Irrigation*, Bangalore, India: Indian Institute of Management.

Seventh five year plan, Planning Commission. (1985). New Delhi: Government of India.

Shah, K. B. (1988). Water cooperatives for water distribution below outlets. In J. S. Kanwar (Ed.), *National seminar on water management - the key to developing agriculture* (pp. 465-481). New Delhi, India: Agricole Publishing Academy.

Sharma, S. K. (1980). Rural development in India: Programmes, strategies, and perspectives. *Community Development Journal*, 15(1).

Shaw, J. (1987). James Buchanan and public choice economics. *Dialogue*, 77.

Sherbondy, J. E. (1985). *Water and power in Inca Cuzco.* Paper presented at the Annual Meeting of the American Anthropological Association.

Shinn, E., & Freeman, D. M. (1988). *A case study of the Niazbeg distributary in Punjab, Pakistan: Linking main and farm irrigation systems in order to control water* (Water Management Synthesis Project Report 69, Vol. 2). Fort Collins, CO: Colorado State University.

Simmons, E. (1984). The economics of irrigated agricultural development and performance in Niger (Annex D). In T. Zalla (Ed.), *Niger irrigation subsector assessment.* Niamey, Niger: United States Agency for International Development.

Singh, K. K. (1982, March). *Farmer's participation in irrigation management: The Pochampad experience.* Paper presented at the Gandhian Institute of Studies, Bombay, India.

_____ . (1987a). *Administered peoples participation in irrigation management: Public tubewell 175 L.G.* Hyderabad, India: Administrative Staff College.

_____ . (1987b). *Tribal lift irrigation cooperative societies: The Sadguru water and development foundation.* Hyderabad, India: Administrative Staff College.

_____ ., & Firdausi, A. A. (1988). *A case study of Pani Panchayat in Mula Command.* Hyderabad, India: Administrative Staff College.

Sixth five year plan, Planning Commission. (1985). New Delhi: Government of India.

Siy, R. Y., Jr. (1982). *Community resource management lessons from the Zarjera.* Quezon City: University of the Philippines Press.

Skogerboe, G. V. (1986, May). *Project paper design in Nepal and Sri Lanka.* Paper presented at the Irrigation Systems Research Forum. Cornell University, Ithaca, NY.

Somerville, C. (1986). *Drought and aid in the Sahel.* Boulder, CO: Westview Press.

Sorbo, G. (1985). *Tenants and nomads in eastern Sudan.* Uppsala, Sweden: Scandinavian Institute of African Studies.

Spicer, E. H., (Ed.). (1967). *Human problems in technological change: A casebook.* New York: John Wiley and Sons, Inc.

Sproule-Jones, M. (1982). Public choice theory and natural resources: A methodological explication and critique. *The American Political Science Review, 76,* 790-804.

Ssennyonga, J. W. (1986, May). *Kenya's transition from the Wittfogelian legacy to community-managed irrigation.* Paper presented at the Irrigation Systems Research Forum, Cornell University: Ithaca, NY.

Stegner, W. (1987, Fall). The Function of Aridity. *Wilderness,* pp. 14-18.

Steinberg, D. I. (1984). *Irrigation and AID's experience: A consideration based on evaluations* (USAID Program Evaluation Report No. 8). Washington, DC: United States Agency for International Development.

252

Stock, R. (1978). The impact of the decline of the Hadejia river floods in Hadejia Emirate, In G. J. van Apeldorn (Ed.), *The aftermath of the 1972-74 drought in Nigeria*. Zaria: Ahmadu Bello University, Federal Department of Water Resources and Centre for Social and Economic Research.

Stone, I. (1984). *Canal irrigation in British India: Perspectives on technological change in a peasant economy*. London: Cambridge University Press.

Tiffen, M. (1985). *Land tenure issues in irrigation planning design and management in sub-Saharan Africa* (Working Paper No. 16). London: Overseas Development Institute.

Tirmizi, J., & Parlin B. (Eds.). (1990). *Societal dynamics and irrigation systems*. Islamabad: Government of Pakistan, Pakistan Council of Research in Water Resources.

Toksoz, S. (1981). *An accelerated irrigation and land reclamation program for Kenya* (Development Discussion Paper No. 114). Cambridge, MA: Harvard Institute for International Development.

Turner, F. (1987, Fall). Slowly sinking in the West. *Wilderness*, pp. 47-50.

United States Department of Agriculture. (1980). *Food problems and prospects in sub-Saharan Africa*. Washington, DC: United States Department of Agriculture.

Udall, J. (1987, March/April). Just add water marketing. *Sierra*, 37-42.

Underhill, H. W. (1984). *Small-scale irrigation in Africa in the context of rural development*. Rome: Food and Agricultural Organization of the United Nations.

Uphoff, N. (1985). *Contrasting approaches to water management development in Sri Lanka*. New York: Third World Legal Studies, International Center for Law and Development.

_____ . (1986a). *Improving international irrigation management with farmer participation: Getting the process right*. Boulder, CO: Westview Press.

_____ . (1986b). Activating community capacity for water management: Experience from Gal Oya, Sri Lanka. In D. Korten (Ed.), *Community management: Asian experience and perspectives*. West Hartford, CT: Kumarian Press.

_____ ., Meinzen-Dick, R., & St. Julien, N. (1985). *Getting the process right: Farmer organization and participation in irrigation water management* (Water management synthesis paper). Ithaca, NY: Cornell University.

_____ ., Wickramasinghe, M. L., & Wijayaratna, C. M. (1990). "Optimum" participation in irrigation management: Issues and evidence from Sri Lanka. *Human Organization, 49*(1), 26-40.

Vainio-Mattila, A. (1985). Socio-economic studies on fuelwood production in Bura settlement scheme in eastern Kenya. In A. Hjort (Ed.), *Land management and survival* (pp. 128-137). Uppsala, Sweden: Scandinavian Institute of African Studies.

van Steekelenburg, P., & Zijlstra, G. (1985). *Evaluation of irrigation projects sponsored by the European economic community.* Wageningen: International Institute for Land Reclamation and Improvement.

VanderVelde, E. J. (1980). Local consequences of a large-scale irrigation system in India. In E. W. Coward (Ed.), *Irrigation and agricultural development in Asia* (pp. 291-328). Ithaca, NY: Cornell University Press.

Veeman, T. S. (1978). Water policy and water institutions in northern India: The case of groundwater rights. *Natural Resources Journal,* 18, 569-587.

Vohra, B. B. (1985). *Some issues in water management.* National Seminar on Integrated Command Area Development, Institute of Command Studies, Bangalore, India.

Wade, R. (1979). The social response to irrigation: An Indian case study. *Journal of Development Studies,* 16(1), 9.

_____. (1980). Managing the main systems. *Economic and Political Weekly,* 15(39), 102-112.

_____. (1982a). The system of administrative and political corruption: Canal irrigation in South India. *Journal of Development Studies,* 18, 287-328.

_____. (1982b). The World Bank and India's irrigation reform. *The Journal of Development Studies,* 18(2), 169-180.

_____. (1987). Managing water managers: Expropriation, or equity as a control mechanism. In W. R. Jordan (Ed.), *Water and water policy in world food supplies* (pp. 177-183). College Station, TX: Texas A&M University Press.

_____. (1988). The management of irrigation systems: How to evoke trust and avoid the prisoners' dilemma. *World Development,* 16(4), 489-500.

_____. (1990). On the 'technical' causes of irrigation hoarding behavior, or why irrigators keep interfering in the main system. In R.K. Sampath & R.W. Young, (Eds.) *Social, economic, and institutional issues in third world irrigation management* (pp. 175-193). Boulder, CO: Westview.

Walter, M. F. (1983). *Small-scale irrigation in the Sierra* (Water Management Synthesis Project Trip Report). Ithaca, NY: Cornell University.

Wattenburger, P. (1986). *Pre-rehabilitation diagnostic analysis of the Niazbeg subproject* (Water Management Synthesis Project Report). Fort Collins, CO: Colorado State University.

Weber, M. (1947). *The theory of social and economic organization.* New York: The Free Press.

Weil, C. E. (1983). Appendix G: Impact on women. In J. L. Wilkinson (Ed.), *Improved land and water use in the Sierra: United States Agency for International Development project impact evaluation.* United States Agency for International Development.

Welsh, F. (1985). *How to create a water crisis.* Boulder, CO: Johnson.

Wensley, C.,-& Walter, M. (1985). *Small-scale irrigation: Design issues in government-assisted systems* (Water Management Synthesis Project Report No. 39). Ithaca, NY: Cornell University.

White, G. F. (1984). Water response adequacy: Illusion and reality. In J. L. Simon & H. Kalm (Eds.), *A response to global 2000* (pp. 250-266). New York: Basil Blackwell.

Whyte, W. F., & Boynton D. (Eds.). (1983). *Higher yielding human systems for agriculture.* Ithaca, NY: Cornell University Press.

Wickramasinghe, L. (1982). *The Gal Oya Experiment.* Paper presented at the Workshop on Water Management in Sri Lanka, Agrarian Research and Training Institute, Colombo, Sri Lanka.

Wijayaratne, C. M. (1984). *Involvement of farmers in water management in Gal Oya, Sri Lanka.* Paper prepared for Food and Agriculture Organization/United States Agency for International Development Expert Consultation on Water management, Yogyakarta, Indonesia.

Wilkens-Wells, J., Wilkens-Wells, P., & Freeman, D. M. (1988). Two tank systems in Polonnaruwa district, Sri Lanka. In D. M. Freeman (Ed.), *Linking main and farm irrigation systems in order to control water* (Water Management Synthesis Project Report 69, Vol. 5). Fort Collins, CO: Colorado State University.

Witfogel, K. (1957). *Oriental despotism: A comparative study of total power.* New Haven, CT: Yale University Press.

World Resources Institute and the International Institute for Environment and Development. (1986). *World resources 1986.* New York: Basic Books.

Worster, D. (1987, Fall). An End to ecstasy. *Wilderness*, pp. 18-34.

Zolezzi, O. (1985). *Gal Oya Water Management Project* (Special Study, Agrarian Research and Training Institute). Colombo, Sri Lanka: Agrarian Research and Training Institute.

About the Contributors

Benjamin U. Bagadion was Engineering Director and later Assistant Administrator for Operations of the National Irrigation Administration (NIA) of the Philippines until his retirement from that agency in 1985. As Assistant Administrator he initiated and directed the program on farmers participation in the NIA described in Chapter 8 until its institutionalization in the agency. Currently, he divides his time consulting in the Philippines and other Asian countries for a number of international agencies.

Jeffrey D. Brewer is currently Chief of Party for Louis Berger International Inc. for the USAID funded Highlands Agricultural Development Project in Guatemala. He has previously worked as organizational specialist on irrigation management and agricultural development projects in India, Sri Lanka, Indonesia, and Pakistan.

David M. Freeman is Professor of Sociology at Colorado State University. He has written widely on the social aspects of irrigation development. His most recent research is summarized in *Local Organizations for Social Development: Concepts and Cases of Irrigation Organization* (1989).

W. Randall Ireson is currently a consultant for Community Aid Abroad working on farmer managed irrigation systems in Laos. He has also been involved with irrigation development in Pakistan and Thailand.

Barbara D. Lynch, extension associate in natural resources at Cornell University is also a consultant on international irrigation and environmental issues, particularly in Andean South America. A development sociologist, she is author of *The Bureaucratic Transition: Government Intervention in Small-Scale Irrigation in Highland Peru*, (1988) and various articles on Andean irrigation and has edited several volumes on water management and environment policy.

Jon R. Moris is Professor of Anthropology at Utah State University and author of numerous articles and books on African development, most recently *Irrigation Development in Africa* (1990, with D. Thom) and *Extension Alternatives in Tropical African Development* (1991).

256

Pamela J. Riley is an associate professor of sociology and coordinator for the Women in International Development Program at Utah State University. She also lectures for the International Irrigation Center on farmer participation in irrigation organization. She has most recently conducted research on the impacts of irrigation projects on women in the Dominican Republic and in Lesotho.

K. K. Singh, until recently distinguished professor of management, Administrative Staff College of India, Hyderabad is now with the Institute of Resource Development and Social Management, Hyderabad as a director and is with the School of Business and Management, Temple University as a visiting professor. He has researched and published in the field of irrigation management with a focus on the institutionalization of farmers' participation on government owned irrigation systems in India.

About the Book and Editors

Irrigation projects consistently fail to meet their projected performance in hydrologic, agronomic, or economic terms. Too often the state bureaucracy seeking to develop irrigation places the blame for such failure on the project beneficiaries -- farmers. Citing problems such as water theft, fee non-payment, and the taking of irrigation water out of turn, analysts fail to examine the broader institutional and organizational factors that shape the behavior of farmers. Moreover, little emphasis is given to the lack of accountability of irrigation agency bureaucrats. This book explores the organizational and institutional factors that lead to poor project performance, but more important, it identifies those managerial factors that have repeatedly been shown to contribute to project success. Taking rational choice theory as its starting point, the book elaborates a perspective that can be used to democratize, decentralize, and privatize irrigation organizations. Case examples from around the world are used to illustrate the interrelationships between project performance, irrigation organization, and farmer participation.

Bradley W. Parlin is Professor of Sociology and Associate Director of the Institute for International Rural & Community Development, Utah State University. **Mark W. Lusk** is Professor of Social Work and Director of the Institute for International Rural & Community Development, Utah State University. Both editors have extensive experience in irrigation development in South Asia, Africa and Latin America.